软件工程专业英语

第2版

吕云翔◎编著

清华大学出版社

北京

内 容 简 介

本书是按照最新的《大学英语教学大纲》对专业英语的要求，以一个大学本科二年级学生 Kevin 与他的同学在一个酒店管理信息系统的实际项目中进行专业实践，直至经过求职面试进入一家 IT 企业工作为主线，将 IT 行业中所需的英语"听、说、读、写、译"基本技能与项目从开始到结束的整个流程有机地融合起来。

本书包括 10 个单元，每个单元都分为听与说、读与译以及模拟写作部分。听与说部分描述了软件开发的技术场景；读与译部分给出了与软件工程相关的文章；模拟写作部分则重点介绍如何撰写技术/商务文档和技术报告等。

本书注重听、说、读、写、译能力的全面发展，适用于高等院校软件工程及其相关专业、软件学院、各类职业信息技术学院和专业培训机构等。本书配有教辅资源，包括听与说的录音、为教师提供授课 PPT，可登录清华大学出版社网站 www.tup.com.cn 下载。

本书封面贴有清华大学出版社防伪标签，无标签者不得销售。
版权所有，侵权必究。举报：010-62782989，beiqinquan@tup.tsinghua.edu.cn。

图书在版编目（CIP）数据

软件工程专业英语/吕云翔编著．—2 版．—北京：清华大学出版社，2021.9（2024.12重印）
（清华科技大讲堂丛书）
ISBN 978-7-302-57980-9

Ⅰ.①软… Ⅱ.①吕… Ⅲ.①软件工程-英语-高等学校-教材 Ⅳ.①TP311.5

中国版本图书馆 CIP 数据核字（2021）第 065468 号

策划编辑：魏江江
责任编辑：王冰飞　李　晔
封面设计：刘　键
责任校对：焦丽丽
责任印制：刘海龙

出版发行：清华大学出版社
网　　址：https://www.tup.com.cn，https://www.wqxuetang.com
地　　址：北京清华大学学研大厦 A 座　　邮　编：100084
社　总　机：010-83470000　　邮　购：010-62786544
投稿与读者服务：010-62776969，c-service@tup.tsinghua.edu.cn
质量反馈：010-62772015，zhiliang@tup.tsinghua.edu.cn
课件下载：https://www.tup.com.cn，010-83470236

印　装　者：三河市铭诚印务有限公司
经　　销：全国新华书店
开　　本：185mm×260mm　　印　张：21.75　　字　数：528 千字
版　　次：2014 年 12 月第 1 版　2021 年 10 月第 2 版　　印　次：2024 年 12 月第 5 次印刷
印　　数：24001～25000
定　　价：59.80 元

产品编号：088821-01

Preface 前言
第2版

本书初版《软件工程专业英语》于2014年12月正式出版以来,经过了几次印刷,被许多高校选为"软件工程专业英语"课程的教材,深受师生的喜爱,获得了良好的社会效益。从另外一个角度来看,作者有责任和义务维护好本书的质量,及时更新本书的内容,做到与时俱进。

这些年来,信息技术的发展突飞猛进,在本书初版的内容中已经涉及的一些技术,由于有了进一步的发展,也有必要对其内容做出及时的更新。本次修订改动内容如下。

(1) 在听与说(Listening and Speaking)部分中做了微小的改动,如对所用工具或软件版本的更新,以及对年代的更新。

(2) 在读与译(Reading and Translating)部分中,有关 Section A 的文章,只替换了 Unit 9 中的文章,但替换了 Section B 的所有文章。

(3) 对每一个单元的模拟写作(Simulated Writing)部分都重新进行了梳理,尤其是对样例。

本书配有教辅资源,包括听与说部分的录音、授课 PPT、每个单元各部分内容的翻译,需要的读者可登录清华大学出版社网站 www.tup.com.cn 下载。

本书的作者为吕云翔,曾洪立参与了部分内容的写作以及各种教辅资源的制作。

最后,请读者不吝赐教,及时提出宝贵意见。

吕云翔
2021年10月

Preface 前言 第1版

英语是全球 IT 行业的行业语言，英语技能是 IT 行业最基本的技能之一，因此熟练掌握相关英语技能对于发展职业生涯具有积极的影响。

本书是按照最新的《大学英语教学大纲》对专业英语的要求，为开设"软件工程专业英语"课程而编写的教材。本书在满足软件工程英语教学的同时，注重学生的实际应用与调动学生的学习兴趣。本书选材广泛，内容丰富，涉及软件开发的项目启动、需求获取、项目计划、团队合作、系统设计、系统实现、系统测试和系统交付的完整流程，并包括如何进行面试和在 IT 企业工作的场景。为了拓展知识，还包括了云计算、移动互联网、大数据、物联网、社交网络、基于位置的服务、通用计算图形处理器、比特币、三维打印和谷歌眼镜等深刻影响着人们生活的信息技术。

本书设计以一个大学本科二年级学生 Kevin 与他的同学在一个酒店管理信息系统的实际项目中进行专业实践，直到经过求职面试进入一个 IT 企业工作，成为这个企业的软件从业人员为主线来组织内容，包含在 IT 行业中所需掌握的英语听说读写译基本技能，涉及从接受项目开始到项目开发完毕这样一个完整的软件开发流程中 IT 相关人员所需的英语听说读写译技能。

本书共有 10 个单元，每个单元的训练都分为听与说、读与译、写等方面。听力部分概要讲述与软件工程相关的知识，对话部分涉及实际工作中与同学、客户或同事之间的交流；读与译的部分包括与软件工程相关的文章和与 IT 相关的最新文章；写作部分讲解如何撰写软件工程文档以及商务文档等。本书注重英语听说读写译能力的全面发展，并与软件工程的专业课程紧密结合，采用场景式教学和体验式学习相结合的方式，实用性强。

另外，本书配有教辅资源，包括听与说的录音、授课 PPT，需要的教师可登录清华大学出版社网站 www.tup.com.cn 下载。使用本书时，建议教学时长为 36~48 学时，教师可根据具体情况进行取舍。

本书适用于高等院校软件工程、计算机科学及其相关专业、软件学院、各类职业信息技术学院和专业培训机构等。书中设计的听力、口语、阅读、写作和翻译的场景融合了角色扮演、多人对话、小组讨论等行之有效的训练方法，能较好地满足课堂教学的需要，有利于学生在课堂上的即时消化吸收。

本书在编写过程中得到了美国专家 Eric Langager 和 Law Phew 的指导，以及杨雪的大力帮助，在此表示衷心的感谢。

  本书试图融合听、说、读、写、译各项技能训练，书中难免会有不尽如人意之处，敬请专家与读者不吝赐教，以使本书臻于完善。

编 者

2014年6月于北京

教学建议

本书的教学建议如下。

(1) 听与说部分(Listening and Speaking)。

对话部分(Dialogue)：教师可先让学生听对话录音，并以提问的方式，引导学生根据所听信息概括对话的主要内容，让学生了解和学习对话中涉及的相关知识。然后，教师可将学生分成三人小组，让其中一组或两组(分别)朗读这段对话，并纠正学生的发音；可让一组或两组参照已有对话并通过替换右边栏中的词语或句子，组织完成一段类似的对话，并对学生完成的情况加以点评。

听力理解部分(Listening Comprehension)：教师可先让学生听短文录音和短文后的问题，让学生根据所听内容选择正确的答案。若播放一遍短文学生感觉有难度，教师可酌情增加录音播放次数。教师最后公布答案，并且讲解相应的单词、短语、缩写及句子，解释这篇短文的重点和难点。另外，可让学生读一遍原文。

听写部分(Dictation)：教师根据实际情况播放 1~3 遍短文录音，让学生根据所听内容填空，将短文补充完整。短文填充完整后，教师最后公布答案，并且讲解相应的单词、短语、缩写及句子，解释这篇短文的重点和难点。另外，可让学生读一遍原文。

(2) 读与译部分(Reading & Translating)。

这部分的内容为软件工程领域知识，使读者深入了解和掌握软件工程相关专业知识。教师可让学生阅读文章(教师可根据文章的长短和难易程度来设定阅读的时间)，并完成文章后的练习。之后教师公布练习答案，并讲解文章后的单词表、短语表、缩写词表和复杂句子来帮助学生进一步理解这篇文章。另外，建议教师讲解这篇文章所涉及的软件工程相关知识。如果课堂时间不够，可将 Section B 作为学生课后的作业。

(3) 模拟写作部分(Simulated Writing)。

教师可先让学生阅读写作方法指导，并配合本书的写作样例进行讲解和指导。教师还可根据实际情况设置场景，让学生根据写作指导并参照写作样例完成一篇类似的文章。如果课堂时间不够，教师可建议学生课下自学"模拟写作部分"。

本书各单元的教学内容、学习要点及教学要求以及课时安排参见下表。

教 学 内 容	学习要点及教学要求	课时安排
Unit 1 Starting a Software Project	• 理解软件项目的基本概念、主要特点、组成要素； • 理解软件工程的基本概念、意义及其主要发展历程； • 理解 AI 如何改变软件工程的 5 个领域； • 掌握备忘录的写作方法	4
Unit 2 Capturing the Requirements	• 理解与用户沟通、获取软件需求的过程； • 理解需求工程的基本概念、作用和主要目标； • 理解需求阶段的主要活动和主要方法； • 掌握需求分析的主要方法和最终产品； • 理解在软件项目中客户与最终用户的区别； • 理解用例及场景的概念； • 掌握软件需求规格说明书的写作方法	4～6
Unit 3 Planning the Project	• 理解软件项目计划的基本概念、作用和主要目标； • 理解项目计划的主要活动和主要方法； • 理解软件项目中需要管理的变量要素及其之间的相互作用； • 理解未来 AI 如何驱动项目管理； • 掌握软件项目计划文档的写作方法	3～5
Unit 4 Working in a Team	• 理解软件项目中团队合作的重要性； • 理解软件项目中团队结构的分类、各自特点及适用情况； • 理解敏捷软件开发的基本概念和主要特点； • 理解如何创建高效的软件开发团队； • 掌握 PowerPoint 演讲稿的写作方法	3～5
Unit 5 Designing the System	• 理解软件系统设计的基本概念、作用和主要目标； • 理解软件设计阶段的主要活动和主要方法； • 理解用户体验和用户界面设计的重要性； • 理解 UX 设计与 UI 设计的区别； • 掌握软件设计规格说明书的写作方法	4～6
Unit 6 Implementing the System	• 理解软件系统编码实现的基本概念、作用和主要目标； • 理解编码阶段的主要活动； • 理解编写高质量代码的主要方法； • 理解程序设计语言的主要发展阶段、分类，及其典型代表语言； • 理解软件开发中的"20-80 法则"； • 理解什么是够用的软件； • 掌握进度报告的写作方法	3～6
Unit 7 Testing the System	• 理解软件系统测试的基本概念、作用和主要目标； • 理解软件测试阶段的主要活动和主要测试方法； • 理解冒烟测试的基本概念； • 理解程序员和测试员一起工作的过程； • 掌握软件测试规格说明书的写作方法	4～6

续表

教 学 内 容	学习要点及教学要求	课时安排
Unit 8 Delivering the System	• 理解软件部署的基本概念、作用和主要目标； • 理解软件交付的主要工作； • 理解软件维护的主要工作； • 理解软件缺陷（Bug）和调试缺陷的基础知识； • 理解软件交付管理的重要性； • 掌握用户手册的写作方法	3～6
Unit 9 Taking an Interview	• 熟悉面试基本技巧和常见面试问题； • 理解软件配置管理的基础知识； • 理解软件产品创建和软件平台构建； • 理解安全的软件开发生命周期的概念； • 掌握简历的写作方法	2
Unit 10 Beginning Your Work	• 理解 IT 公司的组织结构、管理层次、各部门职责； • 理解开源运动； • 理解在软件工程中为什么需要解决道德问题； • 掌握商务电子邮件的写作方法	2
教学总学时建议		32～48

说明：

① 本书为软件工程、计算机科学与技术及相关本科专业"软件工程英语"课程教材，理论授课学时数为 32～48 学时，不同专业可根据不同的教学要求和计划教学时数对教材内容进行适当取舍。

② 非软件工程或计算机类本科专业使用本教材可适当降低教学要求。

③ 理论授课学时数（32～48 学时）包含课堂讨论、练习等必要的课内教学环节。

④ 建议授课时间比例为听与说部分 40%、读与译部分 40%、写作部分 20%。

Contents 目录

随书资源

Unit 1　Starting a Software Project 软件项目启动 ················ 001

Part 1　Listening & Speaking ················ 002
　Dialogue：Starting a Software Project ················ 002
　Listening Comprehension：Software Engineering ················ 005
　Dictation：Mythical Man-Month & No Silver Bullet ················ 006
Part 2　Reading & Translating ················ 007
　Section A：Software Engineering ················ 007
　Section B：5 Areas of Software Engineering AI will Transform ················ 012
Part 3　Simulated Writing：Memo ················ 018

Unit 2　Capturing the Requirements 需求获取 ················ 022

Part 1　Listening & Speaking ················ 023
　Dialogue：Communication with Customers ················ 023
　Listening Comprehension：Software Requirements ················ 025
　Dictation：The Difference between Customer and End-User ················ 026
Part 2　Reading & Translating ················ 027
　Section A：Software Requirements ················ 027
　Section B：Use Cases & Scenarios ················ 031
Part 3　Simulated Writing：Software Requirements Specification ················ 037

Unit 3　Planning the Project 项目策划 ················ 070

Part 1　Listening & Speaking ················ 071
　Dialogue：Software Project Planning ················ 071
　Listening Comprehension：Software Project Planning ················ 073
　Dictation：Four Variables in Projects ················ 074
Part 2　Reading & Translating ················ 076
　Section A：Software Project Plan ················ 076
　Section B：AI-powered Project Management in the Near Future ················ 081
Part 3　Simulated Writing：Software Project Plan ················ 085

Unit 4　Working in a Team 团队合作 ·· 103

Part 1　Listening & Speaking ·· 104
　　Dialogue：Team Structure ·· 104
　　Listening Comprehension：Project Team ·· 105
　　Dictation：Agile Software Development ·· 107
Part 2　Reading & Translating ·· 108
　　Section A：Team Structure ·· 108
　　Section B：How to Build a Highly Effective Software
　　　　　　　Development Team ·· 112
Part 3　Simulated Writing：PowerPoint Presentation ·· 117

Unit 5　Designing the System 系统设计 ·· 123

Part 1　Listening & Speaking ·· 124
　　Dialogue：Software Design ·· 124
　　Listening Comprehension：Software Design ·· 125
　　Dictation：User Interface Design ·· 126
Part 2　Reading & Translating ·· 128
　　Section A：Software Design ·· 128
　　Section B：What is the Difference between UX Design and
　　　　　　　UI Design? ·· 133
Part 3　Simulated Writing：Software Design Specification ·· 139

Unit 6　Implementing the System 系统实现 ·· 172

Part 1　Listening & Speaking ·· 173
　　Dialogue：Creating High-Quality Code ·· 173
　　Listening Comprehension：Writing the Code ·· 175
　　Dictation：Concentrate on the Vital Few, Not the Trivial Many ·· 176
Part 2　Reading & Translating ·· 177
　　Section A：Computer Programming ·· 177
　　Section B：Good-Enough Software ·· 184
Part 3　Simulated Writing：Progress Report ·· 188

Unit 7　Testing the System 系统测试 ·· 194

Part 1　Listening & Speaking ·· 195
　　Dialogue：Software Testing ·· 195
　　Listening Comprehension：Software Testing ·· 197
　　Dictation：Smoke Test ·· 198
Part 2　Reading & Translating ·· 199
　　Section A：Software Testing ·· 199

Section B：Testers and Programmers Working Together ········· 204
Part 3　Simulated Writing：Software Test Specification ··················· 209

Unit 8　Delivering the System 系统交付 ·················· 227

Part 1　Listening & Speaking 🔊 ·················· 228
　　Dialogue：Software Deployment ·················· 228
　　Listening Comprehension：Software Delivery ·················· 230
　　Dictation：Bug and Debugging ·················· 231
Part 2　Reading & Translating ·················· 232
　　Section A：Software Maintenance ·················· 232
　　Section B：Why Software Delivery Management Matters ·················· 238
Part 3　Simulated Writing：User Guide ·················· 244

Unit 9　Taking an Interview 参加面试 ·················· 253

Part 1　Listening & Speaking 🔊 ·················· 254
　　Dialogue：Interview ·················· 254
　　Listening Comprehension：Expansion Company ·················· 256
　　Dictation：Software Configuration Management ·················· 257
Part 2　Reading & Translating ·················· 259
　　Section A：Building Software Products vs Platforms ·················· 259
　　Section B：Secure Software Development Life Cycle ·················· 264
Part 3　Simulated Writing：Resume ·················· 269

Unit 10　Beginning Your Work 开始工作 ·················· 274

Part 1　Listening & Speaking 🔊 ·················· 275
　　Dialogue：Beginning Your Work ·················· 275
　　Listening Comprehension：The Organizational Structure of a Company ······ 277
　　Dictation：Open Source Movement ·················· 278
Part 2　Reading & Translating ·················· 280
　　Section A：The Organizational Structure of a Company ·················· 280
　　Section B：Why We Need to Address Ethical Issues in
　　　　　　　Software Engineering ·················· 285
Part 3　Simulated Writing：Business E-mail ·················· 291

Glossary　（词汇表） ·················· 296

Abbreviation　（缩略语表） ·················· 316

Answers　（练习答案） ·················· 318

Unit 1

Starting a Software Project

软件项目启动

Part 1

Listening & Speaking

Unit 1

Dialogue: Starting a Software Project

(*Kevin*, *Sharon*, *and Jason are three sophomores in the school of software in Beihang University*. *Today*, *they are attending a class meeting at the end of the fourth semester before starting the summer vacation*.)

Teacher: Morning, everyone. In this vacation, you will implement a real project as your course project. There are some subjects you can choose in terms of your interests and experience. Please submit your decision to me within the next week.

Kevin: Excuse me, teacher. Is it a single task or can it be a cooperative work?

Teacher: Team work is recommended, because it benefits you to learn how to work together with your colleagues in the future and how to communicate, share, express, and understand ideas as a team member. But the size of the group should not be more than 4 persons.

Sharon: I'm interested in the subject of Four Seasons Hotel Management Information System, what about you, Kevin?

Kevin: Oh, it is my opinion too. And I think we can cooperate. Hi, Jason, would you like to join us? [1]

Jason: Oh, yes, I'd like to very much!

Sharon: Ok, now let's discuss on each person's responsibility.

[1] Replace with:
1. Would you like to cooperate with us?
2. Would you like to collaborate with us?
3. Would you like to work together with us?

Unit 1　Starting a Software Project　软件项目启动

Jason： Kevin is good at organizing and has lots of programming experience, so I think he can be our team leader or project manager, in charge of instructing our team and programming practice.

Sharon： I agree.

Kevin： Thanks for your trust. Ok, I will do my best. Besides coding, I think it is necessary to create a database and implement a suite of user interfaces for our software.

Jason： I am interested in databases and willing to be responsible for database building and management.

Sharon： I like art design, so I think I can do the UI design and document writing for our project.

Kevin： Oh! It seems this is a wonderful team and makes me very confident! Now, let's divide the work according to the phases of the project in general. As the team leader, I will be responsible for requirements, Jason will be in charge of design and Sharon will take charge of testing.

Jason： Next, we can talk over a rough progress plan for our project.

Kevin： We can design, and then accomplish the UI operation according to the original requirements document provided by our teacher first. At the same time, Jason can be building the database. Finally, we can accomplish coding together.

Sharon： It sounds wonderful. But I am afraid that the contents of the original requirements document will not be sufficient for our design.[2] First of all, we must do the requirements analysis based on the original requirements, and complete a formal Software Requirements Specification as our guidance of design.

[2] Replace with: But I am afraid that I have not enough business knowledge about hotel management.

Kevin: Oh, yes. Thanks for your important **reminder**. What do you think about it, Jason?

Jason: I agree with you completely.

(*After meeting, Kevin asked for a document from the teacher about the hotel business requirements.*)

Kevin: Hi, everybody. I have just got the business requirements of the hotel from our teacher.

Jason: Let me see. Oh, there is a list about their daily business and a table of related requirements. But it seems a little rough without enough detailed procedures, I am afraid.

Kevin: I see. And it does not mention the data flow and business model of this hotel.

Sharon: So, in that case, I think we need some communication with the customer (Four Seasons Hotel) to acquire more information.

Kevin: Yes. It's very necessary and I will call the customer to make an appointment with them. Before that, I think there is something we should do. That is, we had better do some homework to learn some knowledge about basic hotel business and management.

Sharon: That's right! It is very necessary to get information about their business, and will be valuable for us to adequately and accurately understand the requirements.

Jason: Ok, I believe that the Internet can help us a lot.

Exercises

Work in a group, and make up a similar conversation by replacing the statements with other expressions on the right side.

Unit 1　Starting a Software Project　软件项目启动

 Words

sophomore['sɔfəmɔː(r)] n. 大学二年级学生
rough[rʌf] adj. 初步的，粗略的

specification[ˌspesifi'keiʃn] n. 说明书，规范
reminder[ri'maində(r)] n. 提醒，提示

 Phrases

in charge of　负责，领导
a suite of　一系列，一套
take charge of　担任，监管
talk over　商议，讨论

 Abbreviations

UI　User Interface　用户界面

Listening Comprehension：Software Engineering

Listen to the article and the following 3 questions based on it. After you hear a question, there will be a break of 15 seconds. During the break, you will decide which one is the best answer among the four choices marked（A），（B），（C）and（D）.

Questions

1. Which is correct about the development of software according to the article?
   （A）It emerged with software engineering at the same time.
   （B）For a half-century development, it has almost solved problems of high-quality, on-time and within-budget.
   （C）It was just a specialized problem solving and information analysis tool in its early years of development.
   （D）The laws which software evolves according to have changed absolutely during its development.

2. Which point does not belong to the characteristics of software according to the article?
   （A）Easy to change the requirements.
   （B）Easy to adapt the requirement changes.
   （C）Difficult to measure the progress and process of creating

(D) Difficult to test the correctness exhaustively

3. Where was the phrase "software engineering" first used in 1968?
   (A) In a conference
   (B) In a thesis
   (C) In a journal
   (D) In a magazine

 Words

demonstrate[ˈdemənstreit] v. 证明,论证
practice[ˈpræktis] n. 实践,通常的做法,惯例
assimilate[əˈsiməleit] v. 透彻理解,消化
primarily[praiˈmerəli] adv. 原来,根本上
exhaustive[igˈzɔːstiv] adj. 详尽的,彻底的,全面的

address[əˈdres] v. 处理,满足,论述,重点提出
law[lɔː] n. 规则,法则
framework[ˈfreimwɜːk] n. 构架,体系结构,准则
prototype[ˈprəutətaip] n. 原型
discipline[ˈdisəplin] n. 学科,方法

 Abbreviations

NATO   North Atlantic Treaty Organization   北大西洋公约组织

Dictation: Mythical Man-Month & No Silver Bullet

This article will be played three times. Listen carefully, and fill in the numbered spaces with the appropriate words you have heard.

Frederick P. Brooks, Jr., is a Professor of Computer Science at the University of North Carolina at Chapel Hill. He is best ___1___ as the "father of the IBM System/360," having served as ___2___ for its development and later as a manager of the ___3___/360 software project during its design phase.

His book, **Mythical** Man-Month, is a most classic book on the ___4___ elements of software engineering. Since the first ___5___ in 1975, no software engineer's ___6___ has been complete without it. It was in this book that Brooks made the now-famous ___7___ : "Adding ___8___ to a late software project makes it ___9___ ." This has since come to be known as "Brooks's ___10___ ." Software tools and development ___11___ may have changed in the 40 years since the first edition of this book, but the **peculiarly** nonlinear economies of scale in ___12___ work and the nature of ___13___ and groups has not changed an **epsilon**.

Unit 1 Starting a Software Project 软件项目启动

In addition，Brooks is known for No Silver Bullet，which was ____14____ a 1986 IFIPS paper，reprinted in 1987 in the IEEE Computer magazine and ____15____ in the second edition of The Mythical Man-Month later. Silver bullet is used to compare something to make software costs ____16____ as rapidly as computer hardware costs do. "No Silver Bullet" had wide ____17____ and proved provocative. It predicted that a decade would not see any ____18____ technique that would by itself bring an order of magnitude improvement in software ____19____. The author's prediction seems safe. "No Silver Bullet" has ____20____ more and more spirited discussion in the literature than has The Mythical Man-Month.

Words

mythical [ˈmɪθɪkl] *adj.* 神话的，虚构的 peculiarly [pɪˈkjuːliəli] *adv.* 特有地，特别地 epsilon [ˈepsɪlən, epˈsaɪlən] *n.* 小（或近于零）的正数	provocative [prəˈvɒkətɪv] *n.* 引起争论（议论，兴趣等）的 spirited [ˈspɪrɪtɪd] *adj.* 热烈的 literature [ˈlɪtrətʃə(r)] *n.* 著作，文献

Phrases

order of magnitude 数量级

Abbreviations

IFIPS   International Federation of Information Processing Societies   国际信息处理学会联合会
IEEE   Institute of Electrical and Electronics Engineers   美国电气和电子工程师协会

Part 2

Reading & Translating

Section A: Software Engineering

Virtually all countries now depend on complex computer-based systems. National infrastructures and utilities rely on computer-based systems and most electrical products include a computer and controlling software. Industrial manufacturing and **distribution** is completely computerized, as is the financial system. Therefore, producing and maintaining software **cost-effectively** is essential for the functioning of national and international economies.

Software engineering is an engineering discipline whose focus is the cost-effective development of high-quality software systems. Software is abstract and intangible. It is not constrained by materials or governed by physical laws or by manufacturing processes. In some ways, this simplifies software engineering as there are no physical limitations on the potential of software. However, this lack of natural constraints means that software can easily become extremely complex and hence very difficult to understand.

The **notion** of software engineering was first proposed in 1968 at a conference held to discuss what was then called the "software crisis". This software crisis **resulted** directly **from** the introduction of new computer hardware based on **integrated circuits**. Their power made **hitherto** unrealizable computer applications a feasible **proposition**. The resulting software was orders of magnitude larger and more complex than previous software systems.

Early experience in building these systems showed that informal software development was not good enough. **Major** projects were sometimes years late. The software cost much more than predicted, was unreliable, was difficult to maintain and performed poorly. Software development was in crisis. Hardware costs were **tumbling** whilst software costs were rising rapidly. New techniques and methods were needed to control the complexity inherent in large software systems.

These techniques have become part of software engineering and are now widely used. However, as our ability to produce software has increased, so has the complexity of the software systems that we need. New technologies resulting from the **convergence** of computers and communication systems and complex graphical user interfaces **place** new demands **on** software engineers. As many companies still do not apply software engineering techniques effectively, too many projects still produce software that is unreliable, delivered late and over **budget**.

We have made tremendous progress since 1968 and that the development of software engineering has **markedly** improved our software. We have a much better understanding of the activities involved in software development. We have developed effective methods of software specification, design and implementation (Figure 1-1). New **notations** and tools reduce the effort required to produce large and complex systems.

We know now that there is no single "ideal approach" to software engineering. The wide diversity of different types of systems and organizations that use these systems means that we need a diversity of approaches to software development. However, fundamental notions of process and system organization **underlie** all of these techniques,

Unit 1　Starting a Software Project　软件项目启动

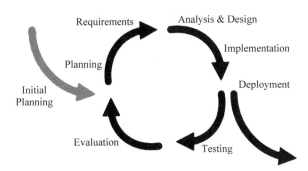

Figure 1-1　Iterative Model of Software Engineering

and these are the essence of software engineering.

　　Software engineers can be rightly proud of their achievements. Without complex software we would not have explored space, would not have the Internet and modern telecommunications, and all forms of travel would be more dangerous and expensive. Software engineering has contributed a great deal, and as the discipline matures, its contributions in this century will be even greater.

 Words

virtually['vɜːtʃuəli] adv. 事实上,实质上	tumble['tʌmbl] v. 使倒下,搅乱
utility[juːˈtiləti] n. 公用事业,公共事业设备	convergence[kənˈvɜːdʒəns] n. 一体化,集中,收敛
distribution[ˌdistriˈbjuːʃn] n. 经销,分销	markedly[ˈmɑːkidli] adv. 显著地,明显地
cost-effective 有成本效益的,划算的	notation[nəuˈteiʃn] n. 符号
notion[ˈnəuʃn] n. 概念,观念,看法	underlie[ˌʌndəˈlai] v. 构成……的基础,位于……之下
hitherto[ˌhiðəˈtuː] adv. 迄今,至今	essence[ˈesns] n. 本质,实质
proposition[ˌprɔpəˈziʃn] n. 主张,提议,建议	rightly[ˈraitli] adv. 确实地
major[ˈmeidʒə] adj. 较大的,较重要的	

 Phrases

result from　由……引起
integrated circuits　集成电路
place on　寄托,把……放在……上
over budget　超过预算

009

Exercises

Ⅰ. Read the following statements carefully, and decide whether they are true (T) or false (F) according to the text.

____ 1. The focus of software engineering is the rapid development of complex software systems.

____ 2. The notion of software engineering was first proposed in a paper in 1968.

____ 3. This software crisis resulted directly from the development of computer hardware.

____ 4. New notations and tools contribute to higher efficiency and less workload in producing large and complex software systems.

____ 5. As an engineering discipline, software engineering has matured adequately today.

Ⅱ. Choose the best answer to each of the following questions according to the text.

1. Which of the following descriptions is not the characteristic of software?
   (A) Abstract and intangible
   (B) Not constrained by materials
   (C) Not governed by physical laws or by manufacturing processes
   (D) Easy to understand and simple to produce as there are no physical limitations

2. What problem(s) existed widely in informal software development in the early years?
   (A) Over schedule
   (B) Cost much more than budget
   (C) Difficult to maintain
   (D) All of the above

3. Which of the following statements is wrong about the techniques in software engineering?
   (A) Techniques are needed to control the complexity of the large software systems.
   (B) Techniques are the essence of software engineering.
   (C) Techniques are now widely used in software engineering.
   (D) New technologies bring new challenges to software engineers continually.

Unit 1 Starting a Software Project 软件项目启动

Ⅲ. Fill in the numbered spaces with the words or phrases chosen from the box. Change the forms where necessary.

> developer   connect   behind   need   involve
> collaborate   cycle   begin   use   refer

Software Engineering vs. Software Development

The difference between software engineering and software development ___1___ with job function. A software engineer may be ___2___ with software development, but few software ___3___ are engineers.

To explain, software engineering ___4___ to the application of engineering principles to create software. Software engineers participate in the software development life cycle through ___5___ the client's needs with applicable technology solutions. Thus, they systematically develop processes to provide specific functions. In the end, software engineering means ___6___ engineering concepts to develop software.

On the other hand, software developers are the driving creative force ___7___ programs. Software developers are responsible for the entire development process. They are the ones who ___8___ with the client to create a theoretical design. They then have computer programmers create the code ___9___ to run the software properly. Computer programmers will test and fix problems together with software developers. Software developers provide project leadership and technical guidance along every stage of the software development life ___10___.

Ⅳ. Translate the following passages into Chinese.

Software Evolution

In software engineering, software evolution is referred to as the process of developing, maintaining and updating software for various reasons. Software changes are inevitable because there are many factors that change during the life cycle of a piece of software. Some of these factors include:

- Requirement changes
- Environment changes
- Errors or security breaches
- New equipment added or removed
- Improvements to the system

For many companies, one of their largest investments in their business is for software and software development. Software is considered a very critical asset and management wants to ensure they employ a team of software engineers who are devoted to ensuring that the software system stays up-to-date with ever evolving changes.

Section B: 5 Areas of Software Engineering AI will Transform

The 5 major **spheres** of software development—software design, software testing, GUI testing, strategic decision making, and automated code generation—are all areas where AI can help. A majority of interest in applying AI to software development is already seen in automated testing and bug detection tools. **Next in line** are the software design **precepts**, decision-making strategies, and finally automating software **deployment pipelines**.

Let's take an **in-depth** look into the areas of high and medium interest of software engineering impacted by AI according to the Forrester Research report (Figure 1-2).

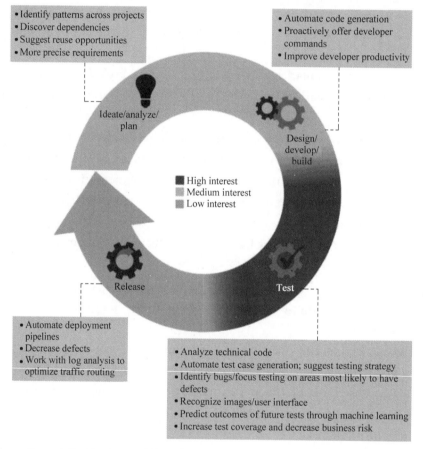

Figure 1-2　The Areas of High and Medium Interest of Software Engineering Impacted by AI According to the Forrester Research Report

1. Software design

In software engineering, planning a project and designing it **from scratch** need designers to apply their specialized learning and experience to come up with alternative solutions before **settling on** a definite solution.

A designer begins with a **vision** of the solution, and after that **retracts** and forwards investigating plan changes until they reach the desired solution. Settling on the correct plan choices for each stage is a tedious and mistake-prone action for designers.

Along this line, a few AI developments have demonstrated the advantages of enhancing traditional methods with intelligent specialists. The catch here is that the operator behaves like an individual partner to the client. This associate should have the capacity to offer opportune direction on the most proficient method to do design projects.

For instance, take the example of AIDA—the Artificial Intelligence Design Assistant, deployed by Bookmark (a website building platform). Using AI, AIDA understands a user's needs and desires and uses this knowledge to create an appropriate website for the user. It makes selections from millions of combinations to create a website style, focus, image and more that are customized for the user. In about 2 minutes, AIDA designs the first version of the website, and from that point it becomes a drag and drop operation.

2. Software testing

Applications interact with each other through countless APIs. They leverage legacy systems and grow in complexity every day. Increase in complexity also leads to its fair share of challenges that can be overcome by machine-based intelligence. AI tools can be used to create test information, explore information authenticity, advancement and examination of the scope and also for test management.

Artificial intelligence, trained right, can ensure the testing performed is error free. Testers freed from repetitive manual tests thus have more time to create new automated software tests with sophisticated features. Also, if software tests are repeated every time source code is modified, repeating those tests can be not only time-consuming but extremely costly. AI comes to the rescue once again by automating the testing for you!

With AI automated testing, one can increase the overall scope of tests leading to an overall improvement of software quality.

Take, for instance, the Functionize tool. It enables users to test fast and release faster with AI enabled cloud testing. The users just have to type a test plan in English and it will be automatically get converted into a functional test case. The tool allows one to elastically scale functional, load, and performance tests across every browser and device in the cloud. It also includes self-healing tests that update autonomously in real-time.

SapFix is another AI Hybrid tool deployed by Facebook which can automatically generate fixes for specific bugs identified by "Sapienz"[1]. It then proposes these fixes to engineers for approval and deployment to production.

3. GUI testing

Graphical User Interfaces (GUI) has become important in interacting with today's software. They are increasingly being used in critical systems and testing them is necessary to avert failures. With very few tools and techniques available to aid in the testing process, testing GUIs is difficult.

Currently used GUI testing methods are ad hoc. They require the test designer to perform humongous tasks like manually developing test cases, identifying the conditions to check during test execution, determining when to check these conditions, and finally evaluate whether the GUI software is adequately tested. Phew! Now that is a lot of work.

Also, not forgetting that if the GUI is modified after being tested, the test designer must change the test suite and perform re-testing. As a result, GUI testing today is resource intensive and it is difficult to determine if the testing is adequate.

Applitools is a GUI tester tool empowered by AI. The Applitools Eyes SDK automatically tests whether visual code is functioning properly or not. Applitools enables users to test their visual code just as thoroughly as their functional UI code to ensure that the visual look of the application is as you expect it to be. Users can test how their application looks in multiple screen layouts to ensure that they all fit the design.

It allows users to keep track of both the webpage behavior, as well as the look of the webpage. Users can test everything they develop from the functional behavior of their application to its visual look.

4. Using Artificial Intelligence in Strategic Decision-Making

Normally, developers have to go through a long process to decide what features to include in a product. However, machine learning AI solution trained on business factors and past development projects can analyze the performance of existing applications and help both teams of engineers and business stakeholders like project managers to find solutions to maximize impact and cut risk.

Normally, the transformation of business requirements into technology specifications requires a significant timeline for planning. Machine learning can help software development companies to speed up the process, deliver the product in lesser time, and increase revenue within a short span.

AI Canvas is a well-known tool for strategic decision-making. The Canvas helps identify the key questions and feasibility challenges associated with building and deploying machine learning models in the enterprise.

The AI Canvas is a simple tool that helps enterprises organize what they need to know into seven categories, namely—Prediction, Judgment, Action, Outcome, Input, Training and feedback. Clarifying these seven factors for each critical decision throughout the organization will help in identifying opportunities for AIs to either reduce costs or enhance performance.

Unit 1　Starting a Software Project　软件项目启动

5. Automatic Code Generation/Intelligent Programming Assistants

Coding a huge project from scratch is often labor intensive and time consuming. An intelligent AI programming assistant will reduce the workload by a great extent.

To combat the issues of time and money constraints, researchers have tried to build systems that can write code before, but the problem is that these methods aren't that good with ambiguity. Hence, a lot of details are needed about what the target program aims at doing, and writing down these details can be as much work as just writing the code. With AI, the story can be flipped.

"Bayou"—an AI—based application is an intelligent programming assistant. It began as an initiative aimed at extracting knowledge from online source code repositories like GitHub. Bayou follows a method called neural sketch learning. It trains an artificial neural network to recognize high-level patterns in hundreds of thousands of Java programs. It does this by creating a "sketch" for each program it reads and then associates this sketch with the "intent" that lies behind the program. This DARPA initiative aims at making programming easier and less error prone.

6. Summing it all up

Software engineering has seen massive transformation over the past few years. AI and software intelligence tools aim to make software development easier and more reliable. According to a Forrester Research report on AI's impact on software development, automated testing and bug detection tools use AI the most to improve software development.

It will be interesting to see the future developments in software engineering empowered with AI. It's expected to have faster, more efficient, more effective, and less costly software development cycles while engineers and other development personnel focus on bettering their skills to make advanced use of AI in their processes.

Words

sphere[sfɪə(r)] n. 范围	associate[əˈsəʊsieɪt, əˈsəʊsiət] n. 同事,伙伴
precept[ˈpriːsept] n. 规则	opportune[ˈɒpətjuːn] adj. 恰好的,适当的
in-depth 深入详尽的,彻底的	proficient[prəˈfɪʃnt] adj. 娴熟的,精通的,训练有素的
vision[ˈvɪʒn] n. 愿景,视野,想象	
retract[rɪˈtrækt] v. 撤回,收回(协议、承诺等)	desire[dɪˈzaɪə(r)] n. & v. 愿望,欲望,渴望
catch[kætʃ] n. 隐藏的困难,暗藏的不利因素	

leverage[ˈliːvərɪdʒ] v. 利用，施加影响 right[raɪt] adv. 正确地，恰当地，彻底地 self-healing 自愈 fix[fɪks] n. 仓促的解决办法 avert[əˈvɜːt] v. 避免，防止 humongous[hjuːˈmʌŋɡəs] adj. 巨大无比的，极大的 phew[fjuː] int. 唉呀，唷（表示不快、惊讶的声音）	empower[ɪmˈpaʊə(r)] v. 使能够，授权 stakeholder[ˈsteɪkhəʊldə(r)] n. 利益相关者，干系者 namely[ˈneɪmli] adv. 即，也就是 flip[flɪp] v. 翻转，快速翻动 repository[rɪˈpɒzətri] n. 知识库，仓库，存储库 better[ˈbetə(r)] v. 改善，胜过，超过

 Phrases

next in line　（按顺序的）下一个
deployment pipeline　部署流水线
from scratch　从头做起
settle on　选定，决定
free from　免于，使摆脱
ad hoc　特别的，专门的
go through　经受，仔细检查
labor intensive　劳动密集型，人工密集
sum up　总结，概述

 Abbreviations

API　Application Programming Interface　应用程序接口
SDK　Software Development Kit　软件开发工具包
DARPA　Defense Advanced Research Projects Agency　美国国防部高级研究计划署

 Notes

[1]　Sapienz 是 Facebook 的 Crash 动态扫描工具。

 Exercises

Ⅰ. Read the following statements carefully, and decide whether they are true（T）or false（F）according to the text.

　　____1. "Bayou" is an intelligent requirement assistant.
　　____2. Coding a huge project from scratch is often very easy.

Unit 1 Starting a Software Project 软件项目启动

_____ 3. The test designer must change the test suite and perform re-testing if the GUI is modified after being tested.

_____ 4. Applitools is another AI Hybrid tool deployed by Facebook.

_____ 5. AI Canvas is a well-known tool for programming.

Ⅱ. **Choose the best answer to each of the following questions according to the text.**

1. How many areas of software engineering will AI transform?
   (A) One
   (B) Three
   (C) Five
   (D) Seven

2. Which of the following statements is an intelligent programming assistant?
   (A) SapFix
   (B) Functionize
   (C) Applitools
   (D) Bayou

3. Which of the following description is right?
   (A) The test designer must change the test suite and perform re-testing if the GUI is modified after being tested.
   (B) Bayou follows a method called neural sketch learning.
   (C) Applitools is a GUI tester tool empowered by AI.
   (D) All of the above.

Ⅲ. **Fill in the numbered spaces with the words or phrases chosen from the box. Change the forms where necessary.**

> aim leader core learn evolve
> out go corner create level

Transformation to Modern Software Engineering

"Transformation"—a thorough or dramatic change in form or appearance—has been __1__ on as long as society itself. AI, blockchain and other innovations of today are, at a basic __2__, just modern iterations of the stone, copper and iron tools our forebears put to work millennia ago. In software engineering we're very much in the throes of our own transformation. For businesses __3__ to grow and prosper in the digital economy, the stakes are similar: get it right, adapt or die.

Accelerating technology advancements over the past two decades ___4___ an environment of continuous disruption. Business ___5___ at big, established enterprises can't afford complacency. Nimble, digital upstarts are lurking right around the ___6___, and many large companies struggle to keep pace.

Robust software is at the ___7___ of the agile, digital business, which means we have an existential imperative to figure this ___8___ and tackle several pressing questions: How is software engineering ___9___ toward "modern" software engineering? What have we ___10___ from others' attempts at transformation? What are the major challenges and risks?

Ⅳ. Translate the following passage into Chinese.

AI-based Approaches for the Management of Complex Software Projects

Artificial Intelligence (AI) has a fundamental influence on all areas of economy, administration and society. An unexpected application of AI lies in software engineering: for the first time, AI provides robust approaches for software development in order to analyze and evaluate complex software and its development processes. Repository Mining, Machine Learning, Big Data Analytics and Software Visualization enable targeted insights and powerful predictions for software quality, software development and software project management.

Part 3

Simulated Writing: Memo

This guide will help you solve your memo-writing problems by discussing what a memo is, describing the parts of memos, and providing examples and explanations that will make your memos more effective.

Audience and Purpose

Memos have a twofold purpose: they bring attention to problems and they solve problems. They accomplish their goals by informing the reader about new information like policy changes, price increases, or by persuading the reader to take an action, such as attend a meeting, or change a current production procedure. Regardless of the specific goal, memos are most effective when they connect the purpose of the writer with the interests and needs of the reader.

Choose the audience of the memo wisely. Ensure that all of the people that the memo is addressed to need to read the memo. If it is an issue involving only one person, do not send the memo to the entire office. Also, be certain that material is not too sensitive to put in a memo; sometimes the best forms of communication are face-to-face interaction or a phone call. Memos are most effectively used when sent to a small to moderate amount of people to communicate company or job objectives.

Parts of a Memo

Standard memos are divided into segments to organize the information and to help achieve the writer's purpose.

- **Heading**

The heading segment follows this general format:
TO: (readers' names and job titles)
FROM: (your name and job title)
DATE: (complete and current date)
SUBJECT: (what the memo is about, highlighted in some way)

Make sure you address the reader by his or her correct name and job title. And be specific and concise in your subject line.

- **Opening**

The purpose of a memo is usually found in the opening paragraph and includes: the purpose of the memo, the context and problem, and the specific assignment or task. Before indulging the reader with details and the context, give the reader a brief overview of what the memo will be about. Choosing how specific your introduction will depend on your memo plan style. The more direct the memo plan, the more explicit the introduction should be. Including the purpose of the memo will help clarify the reason the audience should read this document. The introduction should be brief, and should be approximately the length of a short paragraph.

- **Content**

The context is the event, circumstance, or background of the problem you are solving. You may use a paragraph or a few sentences to establish the background and state the problem. Include only what your reader needs, and be sure it is clear.

- **Task**

One essential portion of a memo is the task statement where you should describe what you are doing to help to solve the problem. Include only as much information as is needed by the decision-makers in the context, but be convincing that a real problem exists. Do no ramble on with insignificant details. If you are having trouble putting the task into words, consider whether you have clarified the situation. You may need to do more planning before you're ready to write your memo. Make sure your purpose-statement forecast divides your subject into the most important topics that the decision-maker needs.

- **Summary**

If your memo is longer than a page, you may want to include a separate summary segment. However, this section is not necessary for short memos and should not take up a significant amount of space. This segment provides a brief statement of the key recommendations you have reached. These will help your reader understand the key points of the memo immediately. This segment may also include references to methods and sources you have used in your research.

- **Discussion**

The discussion segments are the longest portions of the memo, and are the parts in which you include all the details that support your ideas. Begin with the information that is most important. This may mean that you will start with key findings or recommendations. Start with your most general information and move to your specific or supporting facts. (Be sure to use the same format when including details: strongest to weakest.) The discussion segments include the supporting ideas, facts, and research that back up your argument in the memo. Include strong points and evidence to persuade the reader to follow your recommended actions. If this section is inadequate, the memo will not be as effective as it could be.

- **Closing**

After the reader has absorbed all of your information, you want to close with a courteous ending that states what action you want your reader to take. Make sure you consider how the reader will benefit from the desired actions and how you can make those actions easier.

- **Necessary Attachments**

Make sure you document your findings or provide detailed information whenever necessary. You can do this by attaching lists, graphs, tables, etc. at the end of your memo. Be sure to refer to your attachments in your memo and add a notation about what is attached below your closing.

Format

The format of a memo follows the general guidelines of business writing: A memo is usually a page or two long, should be single spaced and left justified. Instead of using indentations to show new paragraphs, skip a line between sentences. Business materials should be concise and easy to read.

You can help your reader understand your memo better by using headings for the summary and the discussion segments that follow it. Write headings that are short but that clarify the content of the segment. The major headings you choose are the ones that should be incorporated in your purpose-statement in the opening paragraph.

For easy reading, put important points or details into lists rather than paragraphs when possible. This will draw the readers' attention to the section and help the audience remember the information better. Using lists will help you be concise when writing a memo.

P. S. **This guide can be tailored according to the different memos.**

A Sample Memo

TO: All Lab Staff
FROM: John Smith, Lab Research Assistant
DATE: August 3, 2020
SUBJECT: New Carpet for Lab

  This weekend, August 8-9, we will be re-carpeting the entire lab room. Therefore, before you leave on Friday, please make sure all wastebaskets, chairs, boxes and other items on the floor are moved into the hallway.

  Also, clear your desktops and either your belongings in your desk drawers (which should be locked) or in boxes. Please label your boxes so that you can find your belongings easily in the next Monday morning.

Kind regards,

John Smith
Lab Research Assistant

 **Exercises**

There are two scenarios as follows, please choose one and write a memo.
Scenario 1: In a team meeting, team members discussed the purchase of a coffee maker for the department, and you should report to your Department Manager, Mr. Wood, about your suggestion and the related research about this affair.
Scenario 2: In the weekly regular meeting, team members planned to hold a party for the New Year, and you should write a memo for this planning, including time, place, assignments, and so on.

Unit 2

Capturing the Requirements

需求获取

Unit 2 Capturing the Requirements 需求获取

Part 1

Listening & Speaking

Unit 2

Dialogue: Communication with Customers

(*Kevin, Sharon and Jason come to the Four Seasons Hotel and have a meeting with Mr. White, the business manager of the hotel and the representative of the end user.*)

Mr. White: Welcome to our hotel.

Kevin: Thank you for giving us the important information about your hotel management needs in your business requirements document. But I'm afraid that there are still some questions that we need to be clear about. In order that we can accomplish the system according to your meanings consistently, we would like to ask you a few questions.

Mr. White: Sure.

Sharon: Mr. White, I am **drafting** the specifications for the Four Seasons Hotel Management Information System, but I found that there may be an important function that is not clear.

Mr. White: Well, what's that?

Sharon: In the case that a customer **books** a room in your hotel, but doesn't **check in** on time, how do you **refund** his **deposit**, the whole, part or none? [1] It will relate to how to depict and **model** the **scenario** of the deposit refund and how to design the module.

Kevin: Yes. It is indeed significant for us to define a clear deposit refund mechanism in the system, because it will not only determine the design of the module itself, but also relate to other modules such as the

[1] Replace with: In the case that a customer books a dinner in your hotel, but doesn't check in on time, how do you deal with it?

Kevin: total amount of the charge. After defining the work flow exactly, we will be able to know how to **deal with** it in the system accordingly.

Mr. White: Well, I see. It is indeed a necessary item in the room booking service. Now, let me explain to you our way in that situation. Usually, if a customer cancels his booking **up to** 24 hours before check-in, we will refund the deposit; and up to 12 hours before check-in, we will refund half, but if the cancellation comes within 12 hours, we don't refund anything.

Kevin: Ok. Well, besides that, how is the rate of the refund deposit determined? Is there any difference between the **VIP** and common customers, or between the **high season** and **low season**?

Mr. White: In our hotel, there is no difference on the rate between the VIP and common customers, but we have **distinctive criteria** in different seasons. From May to October of every year, because of the high season, the timeline for booking cancellation is moved forward, 12 hours earlier than the low season which is from this November to next April. The customer must cancel his booking up to 36 hours before check-in if he determines and wants the whole deposit refunded. If the cancellation is between 36 and 24 hours prior to expected arrival, the refund would be half, and in case of within 24 hours, no refund.

Kevin: Yes, I've got it. And now, it seems clearer and more detailed. We will finish the requirements specification in the next three days and then send it to you this Friday by E-mail.

Mr. White: Ok. No problem! Thanks.

Unit 2 Capturing the Requirements 需求获取

 Exercises

Work in a group, and make up a similar conversation by replacing the statements with other expressions on the right side.

 Words

draft[drɑːft] v. 起草,制定
book[buk] v. 登记,预订
refund['riːfʌnd] v. 退还,偿还
deposit[di'pɔzit] n. 押金,预付金
model['mɔdl] v. 建立模型
scenario[sə'nɑːriəu] n. 场景,某一特定情节

distinctive[di'stiŋktiv] adj. 区别性的,特殊的
criterion[krai'tiəriən] n. 标准,准则(复数为 criteria)

 Phrases

check in   登记,记录,报到
deal with 处理
up to   (时间上)一直到
high season 旺季
low season   淡季

 Abbreviations

VIP Very Important Person   重要人物,大人物

Listening Comprehension: Software Requirements

Listen to the article and answer the following 3 questions based on it. After you hear a question, there will be a break of 15 seconds. During the break, you will decide which one is the best answer among the four choices marked (A), (B), (C) and (D).

Questions

1. Which is the correct order of key steps in the software requirements stage according to this article?
   (A) Inception, elicitation, elaboration, negotiation, and validation
   (B) Elicitation, inception, elaboration, negotiation, and validation

(C) Inception, elicitation, negotiation, elaboration, and validation

(D) Elicitation, inception, negotiation, elaboration, and validation

2. In which step are requirements further expanded into an analysis model?

(A) Negotiation

(B) Validation

(C) Elaboration

(D) Specification

3. In which steps should software team work with other stakeholders in the software requirements stage?

(A) Inception, negotiation, specification, and validation

(B) Inception, elicitation, specification and validation

(C) Inception, elicitation, negotiation, and validation

(D) Inception, elaboration, negotiation, and specification

Words

conduct[kən'dʌkt] v. 实施,进行
generic[dʒə'nerik] adj. 一般的,普通的
distinct[di'stiŋkt] adj. 不同的,有区别的
inception[in'sepʃn] n. 开端
elicitation[iˌlisi'teiʃn] n. 导出,启发
elaboration[iˌlæbə'reiʃn] n. 精化

overriding[ˌəuvə'raidiŋ] adj. 首要的,压倒一切的
facilitate[fə'siliteit] v. 使容易,使便利,促进
class[klɑːs] n. 类
reference['refərəns] v. 把……引作参考,引用

Dictation: The Difference between Customer and End-User

This article will be played three times. Listen carefully, and fill in the numbered spaces with the appropriate words you have heard.

Software ___1___ communicate with many different stakeholders, but customers and end-users have the most significant impact on the technical work that ___2___. In some ___3___ the customer and the end-user are one in the ___4___, but for many ___5___, the customer and the end-user are different people, ___6___ for different managers in different business ___7___.

A customer is the person or ___8___ who: (1) ___9___ requested the software to be ___10___; (2) defines ___11___ business objectives for the software;

(3) provides basic product ___12___ ; and (4) coordinates **funding** for the project. In a ___13___ or system business, the customer is often the marketing ___14___ . In an IT environment, the customer might be a business ___15___ or department.

On the contrary, an ___16___ is the person or group who: (1) will ___17___ use the software that is built to ___18___ some business ___19___ , and (2) will define ___20___ details of the software so that the business purposes can be achieved.

Words

funding ['fʌndiŋ] n. 提供资金

Phrases

on the contrary 正相反

Part 2

Reading & Translating

Section A: Software Requirements

The main goal of the requirements phase is to produce the Software Requirements Specification (SRS), which accurately **captures** the client's requirements and which forms the basis of software development and **validation**. The basic reason for the difficulty in **specifying** software requirements comes from the fact that there are three **interested** parties — the client, the end users, and the software developer. The requirements document has to be such that the client and users can understand it easily and the developers can use it as a basis for software development. Due to the diverse parties involved in software requirements specification, a communication gap exists. This makes the task of requirements specification difficult (Figure 2-1).

There are three basic activities in the requirements phase. The first is problem or requirement analysis. The goal of this activity is to understand such different aspects as the requirements of the problem, its **context**, and how it fits within the client's organization. The second activity is requirements specification, during which the understood problem is specified or written, producing the SRS. And the third activity is requirements validation, which is done to ensure that the requirements specified in the SRS are indeed what are desired.

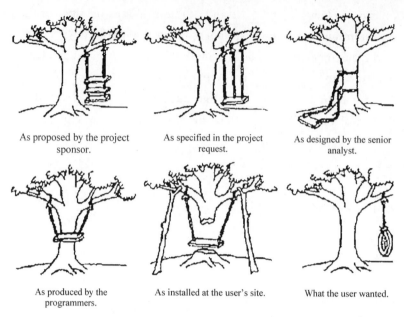

Figure 2-1 Specifying Software Requirements is Difficult

There are three main approaches to analyze. Unstructured approaches rely on interaction between the analyst, customer, and user to reveal all the requirements (which are then **documented**). The second is the modeling-oriented approach, in which a model of the problem is built based on the available information. The model is useful in determining if the understanding is correct and in ensuring that all the requirements have been determined. Modeling may be function-oriented or object-oriented. The third approach is the prototyping approach in which a prototype is built to validate the correctness and completeness of requirements.

To satisfy its goals, a SRS should possess characteristics like completeness, consistency, unambiguous, verifiable, modifiable, etc. A good SRS should specify all the functions the software needs to support, **performance** of the system, the design constraints that exist, and all the external interfaces.

One method for specifying the functional specifications that has become popular is the **use case** approach. With this approach the functionality of the system is specified through use cases, with each use case specifying the behavior of the system when a user interacts with it for achieving some goal. Each use case contains a normal scenario, as well as many exceptional scenarios, thereby providing the complete behavior of the system. Though use cases are meant for specification, as they are natural and story-like, by expressing them at different levels of abstraction they can also be used for problem analysis.

For validation, the most commonly used method is reviewing or inspecting the requirements. In requirements inspections, the team of reviewers also includes a

Unit 2 Capturing the Requirements 需求获取

representative of the client to ensure that all requirements are captured.

Words

capture[ˈkæptʃə(r)] v. 获得,捕获
validation[ˌvæliˈdeiʃən] n. 验证,确认
specify[ˈspesifai] v. 详细说明,指定
interest[ˈintrəst, ˈintrest] v. 使……感兴趣,使……参与
context[ˈkɔntekst] n. 上下文,环境,背景

document[ˈdɔkjumənt] v. 记录,纪实性地描述
performance[pəˈfɔːməns] n. 性能,业绩,工作情况

Phrases

use case 用例

Exercises

Ⅰ. Read the following statements carefully, and decide whether they are true (T) or false (F) according to the text.

_____ 1. Software requirements specification should accurately capture the client's requirements and form the basis of software development and validation.

_____ 2. The goal of requirements validation is to understand such different aspects as the requirements of the problem, its context, and how it fits within the client's organization.

_____ 3. The model and the prototype are both useful for the correctness and completeness of requirements.

_____ 4. The use case approach has become one of the most popular methods for specifying the functional specifications.

_____ 5. In requirements inspections, the representative of the client is responsible for ensuring that all requirements are captured.

Ⅱ. Choose the best answer to each of the following questions according to the text.

1. Which statement is wrong about the SRS?
   (A) It is the most important product in the requirements phase.
   (B) The most direct reason for the difficulty in SRS is the diversity of the parties involved.

(C) A good SRS should specify not only all the functions of the software, but also other non-functional requirements.

(D) SRS should be inspected by the team of reviewers.

2. Which statement is right about the activity of requirement analysis?

(A) Prototype is a theoretic design to validate the correctness and completeness of requirements.

(B) During this activity, the understood problem is specified and written in SRS.

(C) There are three main different approaches for modeling.

(D) The main objective of this activity is to understand the problem.

3. Which statement is wrong about the use case approach?

(A) Use cases specify the functionality of the system.

(B) Use cases can be used for different basic activities in the requirements phase.

(C) Each use case consists of many normal scenarios and many exceptional scenarios.

(D) Use case occurs when a user interacts with the system for achieving some goal.

Ⅲ. **Fill in the numbered spaces with the words or phrases chosen from the box. Change the forms where necessary.**

> expect   face   overcome   say   base
> relate   explain   define   declare   early

Functional Requirement

A functional requirement, in software and systems engineering, is a ___1___ of the intended function of a system and its components. ___2___ on functional requirements, an engineer determines the behavior (output) that a device or software is ___3___ to exhibit in the case of a certain input. A system design is an ___4___ form of a functional requirement.

Functional requirements of a system can ___5___ to hardware, software or both in terms of calculations, technical details, data manipulation and processing or other specific functionality that ___6___ what a system is supposed to accomplish. A functional requirement can be in the form of a document ___7___ the expected types of outputs when the device (system) is placed in a certain kind of environment. A functional requirement is ___8___ to be a later form of a system design because a design is the outcome of ___9___ a certain kind of a problem (technical/non-technical) being ___10___.

Ⅳ. Translate the following passage into Chinese.

Non-functional requirements

Non-functional requirements are sometimes defined in terms of metrics (i. e. something that can be measured about the system) to make them more tangible. Non-functional requirements may also describe aspects of the system that don't relate to its execution, but rather to its evolution over time (e. g. maintainability, extensibility, documentation, etc.).

Non-functional requirements are not straightforward requirements of the system, rather it is related to usability (in some way). For example, for a banking application, a major non-functional requirement will be available where the application should be available 24/7 with no downtime if possible.

Section B: Use Cases & Scenarios

Once you have developed an initial set of Functional Requirements during the Requirements Gathering phase you will have a good understanding of the intended behavior of the system. You will understand what functionality is desired, what constraints are imposed, and what business objectives will be satisfied. However, one shortcoming of a traditional "**laundry-list**" of requirements is that they are static and doesn't concern themselves with the different business processes that need be supported by one feature.

For example, in our **fictitious** online library system, the functionality for managing returns would need to handle the separate situations where a borrower returns a book early and when he/she returns it late. Although the same functionality is involved, they are different situations and the system would need to handle the separate conditions in each use case. Therefore, use-cases are a valuable way of uncovering implied functionality that occurs due to different ways in which the system will be used. Also use-cases provide a great starting point for the **test cases** that will be used to test the system.

A use case is a definition of a specific business objective that the system needs to accomplish. A use-case will define this process by describing the various external **actors** (or entities) that exist outside of the system, together with the specific interactions they have with the system in the accomplishment of the business objective.

Use cases can be described either at an abstract level (known as a business use-case) or at an implementation-specific level (known as a system use case). Each of these is described in more detail below:
- Business Use Case—Also known as an "Abstract-Level Use Case", these use cases are written in a technology—**agnostic** manner, simply referring to the high-level business process being described (e.g. "book return") and the various external

entities (also known as an actor) that take part in the process (e.g. "borrower", "librarian", etc.). The business use case will define the sequence of actions that the business needs to perform to give a meaningful, observable result to the external entity.

- System Use Case—Also known as an "Implementation Use Case", these use cases are written at a lower level of detail than the business use case and refer to specific processes that will be carried out by different parts of the system. For example, a system use case might be "return book when overdue" and would describe the interactions of the various actors (borrower, librarian) with the system in carrying out the end-to-end process.

Typically, you will start by defining the high-level business use-cases, and as the system requirements get defined, they will be "drilled-down" into one or more lower-level system use cases.

One related artifact is the "business scenario" or user story. These are similar to use cases in terms of what they seek to accomplish a description of how the system will carry out a specified business process to fulfill the stated requirements. However, unlike a use case which is a step-by-step enumeration of the tasks carried out during a process (with the associated actors), a scenario is much more free-form.

A user story is typically a narrative that describes how a user would experience the functionality of the system. As the name suggests it is a "short story" that describes the tasks they carry out, what information they see and how they interact with the system. User Stories have become more popular with the advent of Agile Methodologies that emphasize customer collaboration, user interaction and simplicity.

Use case diagrams are diagrams that can be used to more clearly illustrate the set of use cases that are provided by the functionality in a system (Figure 2-2). The diagrams contain both the external entities that will be using the system (also known as "actors") and the discrete use cases (or goals) that the users will be carrying out.

These diagrams are typically represented in the UML modeling language (though other forms do exist) and will help the business analyst convey the relationships between the actors and their business goals and how the design of the system needs to support their different objectives with integrated business processes.

In addition to UML use case diagrams, which depict the different actors and goals, you can use process flow diagrams to enumerate graphically the steps that will take place in each process (Figure 2-3).

In the simplest form, it can be a linear flow from start to finish, but in more complex use cases, you may have multiple branches and decision points in which case it becomes a fully-fledged flowchart.

Use cases are often used as a means of discovering and representing functional and system requirements since they define the interactions and tasks necessary for carrying

Unit 2 Capturing the Requirements 需求获取

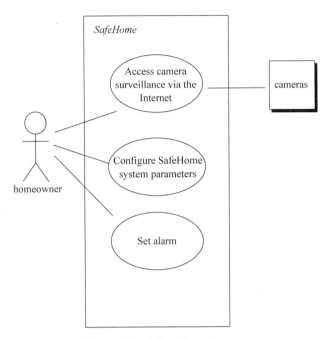

Figure 2-2 A Use Case Diagram

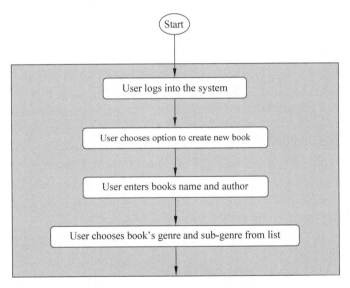

Figure 2-3 A Process Flow Diagram

our specific business objectives. However, they typically are not a good way of defining non-functional requirements such as technical requirements or system qualities. A requirements traceability matrix is used to ensure completeness—namely that all functional requirements are covered by at least one business use case and that all system requirements are covered by at least one system use case.

Since you typically need to ensure that there is complete requirements test coverage for a successful quality assurance program, use cases provide a good starting point for the design of test cases that will be used to test that the system meets the specified requirements.

Once the requirements engineering activities have been completed and the business analysts are happy with the requirements definition, the test writers can create test cases based on the system use cases. This usually involves adding more detailed pre-conditions and post-conditions and writing different test cases "variants" of the same use-case to cover different testing scenarios. Part of this will involve the identification of the critical exception cases that could cause the system to fail, and the development of the necessary test data to ensure full coverage of all test conditions.

Words

laundry-list 细目清单 fictitious [fik'tiʃəs] adj. 虚构的,假想的,编造的 actor ['æktə(r)] n. 角色,行动者 agnostic [æg'nɒstik] adj. 不可知论的,怀疑的 overdue [ˌəʊvə'djuː] adj. 过期的,迟到的 end-to-end 端到端 drill-down 深度探讨 artifact ['ɑːtifækt] n. 人工制品,手工艺品 fulfill [fʊl'fil] v. 实现 enumeration [iˌnjuːmə'reiʃn] n. 列举事实,逐条陈述	free-form 自由形态的,结构不规则的 narrative ['nærətiv] n. & adj. 叙述,故事,讲述的 discrete [di'skriːt] adj. 离散的,不连续的 convey [kən'vei] v. 传达 fully-fledged 完善的,成熟的,羽毛丰满的 pre-condition 前置条件,先决条件 post-condition 后置条件 variant ['veəriənt] n. 变体,转化

Phrases

test case 测试用例
business scenario 业务场景,业务方案
with the advent of 随着……的出现
agile methodologies 敏捷方法论,敏捷开发方法

Abbreviations

UML Unified Modeling Language 统一建模语言

Unit 2 Capturing the Requirements 需求获取

 Exercises

Ⅰ. Read the following statements carefully, and decide whether they are true (T) or false (F) according to the text.

____ 1. Business Use Case is also known as an "Implementation Use Case".

____ 2. System Use Case is also known as an "Abstract-Level Use Case".

____ 3. Use case diagrams contain both the external entities that will be using the system (or goals) and the discrete use cases (also known as "actors") that the users will be carrying out.

____ 4. Use cases typically are not a good way of defining non-functional requirements such as technical requirements or system qualities.

____ 5. Unlike a scenario which is a step-by-step enumeration of the tasks carried out during a process (with the associated actors), a use case is much more free-form.

Ⅱ. Choose the best answer to each of the following questions according to the text.

1. How many levels of use cases can be described?
   (A) One
   (B) Two
   (C) Three
   (D) Four

2. In which of the following languages are use case diagrams typically represented?
   (A) UML
   (B) C++
   (C) Python
   (D) Java

3. Which of the following descriptions is not right?
   (A) A user story is typically a narrative that describes how a user would experience the functionality of the system.
   (B) A use case is a definition of a specific business objective that the system needs to accomplish.
   (C) Use cases typically are not a good way of defining non-functional requirements such as technical requirements or system qualities.
   (D) System Use Case is also known as an "Abstract-Level Use Case".

Ⅲ. Fill in the numbered spaces with the words or phrases chosen from the box.

Change the forms where necessary.

> however   break   only   story   environment
> as   capture   ensure   deliver   through

Managing Requirements in an Agile Environment

In many ways, the manner of ___1___ requirements in an Agile project management environment is similar to a "waterfall," or traditional project management ___2___ —numerous meetings with subject matter experts, end users, walkthrough / documenting the current business workflow, creating mockups, etc. ___3___, Agile and traditional project management approaches contrast in how requirements are managed over time.

Managing requirements in an Agile project management environment is to think ___4___ the full life cycle of the requirement; to consider the full user experience and even beyond the defined stakeholders. Business requirements should be ___5___ down in such a way that supports iterative development and enables flexibility to respond to potential changes as each increment is ___6___ and reviewed by business users and / or customers.

Further, requirements should produce strong, testable user ___7___ that are clarified and reviewed often with the customer, end users, and development team. User stories should promote a consistent conversation with the team that not ___8___ strengthens understanding of the business need, but results in more informed estimation and prioritization. ___9___ both the business users and the development team are engaged throughout the project, it builds trust and encourages collaboration across the board. Finally, it ___10___ faster, better delivery of requirements that truly convey and meet the business need.

Ⅳ. **Translate the following passage into Chinese.**

UML

UML, short for Unified Modeling Language, is a standardized modeling language consisting of an integrated set of diagrams, developed to help system and software developers for specifying, visualizing, constructing, and documenting the artifacts of software systems, as well as for business modeling and other non-software systems. The UML represents a collection of best engineering practices that have proven successful in the modeling of large and complex systems. The UML is a very important part of developing object oriented software and the software development process. The UML uses mostly graphical notations to express the design of software projects. Using the UML helps project teams communicate, explore potential designs, and validate the architectural design of the software.

Part 3

Simulated Writing: Software Requirements Specification

The requirements specification covers exactly the same ground as the requirements definition, but from the perspective of the developers. Where the requirements definition is written in terms of the customer's vocabulary, referring to objects, states, events, and activities in the customer's world, the requirements specification is written in terms of the system's interface. We accomplish this by rewriting the requirements so that they refer only to those real-world objects (states, events, actions) that are sensed or actuated by the proposed system:

(1) In documenting the system's interface, we describe all inputs and outputs in detail, including the sources of inputs, the destinations of outputs, the valve ranges and data formats of input and output data, protocols governing the order in which certain inputs and outputs must be exchanged, window formats and organization, and any timing constraints. Note that the user interface is rarely the only system interface; the system may interact with other software components (e.g. a database), special-purpose hardware, the Internet, and so on.

(2) Next, we restate the required functionality in terms of the interfaces' inputs and outputs. We may use a functional notation or data-flow diagrams to map inputs to outputs, or use logic to document functions' pre-conditions and post-conditions. We may use state machines or events traces to illustrate exact sequences of operations or exact orderings of inputs and outputs. We may use an entity-relationship diagram to collect related activities and operations into classes. In the end, the specification should be complete, meaning that it should specify an output for any feasible sequence of inputs. Thus, we include validity checks on inputs and system responses to exceptional situations, such as violated pre-conditions.

(3) Finally, we devise some fit criteria for each of the customer's quality requirements, so that we can conclusively demonstrate whether our system meets these quality requirements.

The result is a description of what the developers are supposed to produce, written in sufficient detail to distinguish between acceptable and unacceptable solutions, but without saying how the proposed system should be designed or implemented.

Software Requirements Specification Template

Software Requirements Specification
For
< Project Name >

< Version >

< Author >

Prepared for
...
Instructor:
Date

Revision History

Date	Description	Author	Comments
< date >	< Version 1 >	< Your Name >	< First Revision >

Document Approval

The following Software Requirements Specification has been accepted and approved by the following:

Signature	Printed Name	Title	Date
	< Your Name >	Team Leader	
	...	Instructor, ...	

Unit 2 Capturing the Requirements 需求获取

Table of Contents

REVISION HISTORY

DOCUMENT APPROVAL

1. **Introduction**
   - 1.1 Purpose
   - 1.2 Scope
   - 1.3 Definitions, Acronyms, and Abbreviations
   - 1.4 References
   - 1.5 Overview
2. **General Description**
   - 2.1 Product Perspective
   - 2.2 Product Functions
   - 2.3 User Characteristics
   - 2.4 General Constraints
   - 2.5 Assumptions and Dependencies
3. **Specific Requirements**
   - 3.1 External Interface Requirements
     - 3.1.1 User Interfaces
     - 3.1.2 Hardware Interfaces
     - 3.1.3 Software Interfaces
     - 3.1.4 Communications Interfaces
   - 3.2 Functional Requirements
     - 3.2.1 <Functional Requirement or Feature #1>
     - 3.2.2 <Functional Requirement or Feature #2>
   - 3.3 Use Cases
     - 3.3.1 Use Case #1
     - 3.3.2 Use Case #2
   - 3.4 Classes / Objects
     - 3.4.1 <Class / Object #1>
     - 3.4.2 <Class / Object #2>
   - 3.5 Non-Functional Requirements
     - 3.5.1 Performance
     - 3.5.2 Reliability
     - 3.5.3 Availability
     - 3.5.4 Security
     - 3.5.5 Maintainability
     - 3.5.6 Portability
   - 3.6 Inverse Requirements
   - 3.7 Design Constraints

  3.8 Logical Database Requirements
  3.9 Other Requirements
4. **Analysis Models**
  4.1 Sequence Diagrams
  4.2 Data Flow Diagrams (DFD)
  4.3 State-Transition Diagrams (STD)
5. **Change Management Process**
A. **Appendices**
  A.1 Appendix 1
  A.2 Appendix 2

1. Introduction

The introduction to the Software Requirement Specification (SRS) document should provide an overview of the complete SRS document. While writing this document please remember that this document should contain all of the information needed by a software engineer to adequately design and implement the software product described by the requirements listed in this document (Note: the following subsection annotates are largely taken from the IEEE Guide to SRS).

1.1 Purpose

What is the purpose of this SRS and the (intended) audience for which it is written?

1.2 Scope

This subsection should:

(1) Identify the software product(s) to be produced by name; for example, Host DBMS, Report Generator, etc.

(2) Explain what the software product(s) will do, and, if necessary, will not do.

(3) Describe the application of the software being specified. As a portion of this, it should:

(a) Describe all relevant benefits, objectives, and goals as precisely as possible. For example, to say that one goal is to provide effective reporting capabilities is not as good as saying parameter-driven, user-definable reports with a 2h turnaround and on-line entry of user parameters.

(b) Be consistent with similar statements in higher-level specifications (for example, the System Requirement Specification), if they exist. What is the scope of this software product?

1.3 Definitions, Acronyms, and Abbreviations

This subsection should provide the definitions of all terms, acronyms, and abbreviations required to properly interpret the SRS. This information may be provided by reference to one or more appendixes in the SRS or by reference to other documents.

1.4 References

This subsection should:

(1) Provide a complete list of all documents referenced elsewhere in the SRS, or in a separate, specified document.

(2) Identify each document by title, report number—if applicable—date, and publishing organization.

(3) Specify the sources from which the references can be obtained.

This information may be provided by reference to an appendix or to another document.

1.5 Overview

This subsection should:

(1) Describe what the rest of the SRS contains.

(2) Explain how the SRS is organized.

2. General Description

This section of the SRS should describe the general factors that affect the product and its requirements. It should be made clear that this section does not state specific requirements; it only makes those requirements easier to understand.

2.1 Product Perspective

This subsection of the SRS puts the product into perspective with other related products or projects (See the IEEE Guide to SRS for more details).

2.2 Product Functions

This subsection of the SRS should provide a summary of the functions that the software will perform.

2.3 User Characteristics

This subsection of the SRS should describe those general characteristics of the eventual users of the product that will affect the specific requirements (See the IEEE Guide to SRS for more details).

2.4 General Constraints

This subsection of the SRS should provide a general description of any other items that will limit the developer's options for designing the system (See the IEEE Guide to SRS for a partial list of possible general constraints).

2.5 Assumptions and Dependencies

This subsection of the SRS should list each of the factors that affect the requirements stated in the SRS. These factors are not design constraints on the software but are, rather, any changes to them that can affect the requirements in the SRS. For example, an assumption might be that a specific operating system will be available on the hardware designated for the software product. If, in fact, the operating system is not available, the SRS would then have to change accordingly.

3. Specific Requirements

This will be the largest and most important section of the SRS. The customer requirements will be embodied within Section 2, but this section will give the requirements that are used to guide the project's software design, implementation, and testing.

Each requirement in this section should be:

- Correct
- Traceable (both forward and backward to prior/future artifacts)
- Unambiguous
- Verifiable (i.e., testable)
- Prioritized (with respect to importance and/or stability)
- Complete
- Consistent
- Uniquely identifiable (usually via numbering like 3.4.5.6)

Attention should be paid to carefully organize the requirements presented in this section so that they may be easily accessed and understood. Furthermore, this SRS is not the software design document, therefore one should avoid the tendency to over-constrain (and therefore design) the software project within this SRS.

3.1 External Interface Requirements

3.1.1 User Interfaces

3.1.2 Hardware Interfaces

3.1.3 Software Interfaces

3.1.4 Communications Interfaces

3.2 Functional Requirements

This section describes specific features of the software project. If desired, some requirements may be specified in the use-case format and listed in the Use Cases Section.

3.2.1 < Functional Requirement or Feature #1 >

3.2.1.1 Introduction

3.2.1.2 Inputs

3.2.1.3 Processing

3.2.1.4 Outputs

3.2.1.5 Error Handling

3.2.2 < Functional Requirement or Feature #2 >

...

3.3 Use Cases

3.3.1 Use Case #1

3.3.2 Use Case #2

...

3.4 Classes / Objects

3.4.1 < Class / Object #1 >

3.4.1.1 Attributes

3.4.1.2 Functions

< Reference to functional requirements and/or use cases >

3.4.2 < Class / Object #2 >

...

3.5 Non-Functional Requirements

Non-functional requirements may exist for the following attributes. Often these requirements must be achieved at a system-wide level rather than at a unit level. State the requirements in the following sections in measurable terms (e. g., 95% of transaction shall be processed in less than a second, system downtime may not exceed 1 minute per day, > 30 day MTBF value, etc.).

3.5.1 Performance

3.5.2 Reliability

3.5.3 Availability

3.5.4 Security

3.5.5 Maintainability

3.5.6 Portability

3.6 Inverse Requirements

State any useful inverse requirements.

3.7 Design Constraints

Specify design constrains imposed by other standards, company policies, hardware limitation, etc. that will impact this software project.

3.8 Logical Database Requirements

Will a database be used? If so, what logical requirements exist for data formats, storage capabilities, data retention, data integrity, and so on?

3.9 Other Requirements

Catch all sections for any additional requirements.

4. Analysis Models

List all analysis models used in developing specific requirements previously given in this SRS. Each model should include an introduction and a narrative description. Furthermore, each model should be traceable for the SRS's requirements.

4.1 Sequence Diagrams

4.2 Data Flow Diagrams (DFD)

4.3 State-Transition Diagrams (STD)

5. Change Management Process

Identify and describe the process that will be used to update the SRS, as needed, when project scope or requirements change. Who can submit changes and by what means, and how will these changes be approved?

A. Appendices

Appendices may be used to provide additional (and hopefully helpful) information. If present, the SRS should explicitly state whether the information contained within an appendix is to be considered as a part of the SRS's overall set of requirements.

Example Appendices could include (initial) conceptual documents for the software project, marketing materials, minutes of meetings with the customer(s), etc.

    A.1   Appendix 1

    A.2   Appendix 2

P.S. This template can be tailored according to the different projects.

A Sample Software Requirement Specification

Software Requirements Specification
For
Web Publishing System

Version 1.0

Victor Lu (Team leader)
Jack Liang
Jason Wang
Mary Zeng
David Chang

April 10, 2020

Unit 2 Capturing the Requirements 需求获取

Table of Contents

1. **Introduction**
   - 1.1 Purpose
   - 1.2 Scope of Project
   - 1.3 Glossary
   - 1.4 References
   - 1.5 Overview of Document
2. **Overall Description**
   - 2.1 System Environment
   - 2.2 Functional Requirements Specification
     - 2.2.1 Reader Use Case
     - 2.2.2 Author Use Case
     - 2.2.3 Reviewer Use Case
     - 2.2.4 Editor Use Cases
   - 2.3 User Characteristics
   - 2.4 Non-Functional Requirements
3. **Requirements Specification**
   - 3.1 External Interface Requirements
   - 3.2 Functional Requirements
     - 3.2.1 Search Article
     - 3.2.2 Communicate
     - 3.2.3 Add Author
     - 3.2.4 Add Reviewer
     - 3.2.5 Update Person
     - 3.2.6 Update Article Status
     - 3.2.7 Enter Communication
     - 3.2.8 Assign Reviewer
     - 3.2.9 Check Status
     - 3.2.10 Send Communication
     - 3.2.11 Publish Article
     - 3.2.12 Remove Article
   - 3.3 Detailed Non-Functional Requirements
     - 3.3.1 Logical Structure of the Data
     - 3.3.2 Security

1. Introduction

1.1 Purpose

The purpose of this document is to present a detailed description of the Web Publishing System. It will explain the purpose and features of the system, the interfaces of the system, what the system will do, the constraints under which it must operate and how the system will react to external stimuli. This document is intended for both the stakeholders and the developers of the system and will be proposed to the Regional Historical Society for its approval.

1.2 Scope of Project

This software system will be a Web Publishing System for a local editor of a regional historical society. This system will be designed to maximize the editor's productivity by providing tools to assist in automating the article review and publishing process, which would otherwise have to be performed manually. By maximizing the editor's work efficiency and production the system will meet the editor's needs while remaining easy to understand and use.

More specifically, this system is designed to allow an editor to manage and communicate with a group of reviewers and authors to publish articles to a public website. The software will facilitate communication between authors, reviewers, and the editor via E-mail. Preformatted reply forms are used in every stage of the articles' progress through the system to provide a uniform review process; the location of these forms is configurable via the application's maintenance options. The system also contains a relational database containing a list of Authors, Reviewers, and Articles.

1.3 Glossary

Glossary is shown in Table 2-1.

Table 2-1 Glossary

Term	Definition
Active Article	The document that is tracked by the system; it is a narrative that is planned to be posted to the public website
Author	Person submitting an article to be reviewed. In case of multiple authors, this term refers to the principal author, with whom all communication is made
Database	Collection of all the information monitored by this system
Editor	Person who receives articles, sends articles for review, and makes final judgments for publications
Field	A cell within a form
Historical Society Database	The existing membership database (also HS database)
Member	A member of the Historical Society listed in the HS database

Unit 2　Capturing the Requirements　需求获取

continued

Term	Definition
Reader	Anyone visiting the site to read articles
Review	A written recommendation about the appropriateness of an article for publication; may include suggestions for improvement
Reviewer	A person that examines an article and has the ability to recommend approval of the article for publication or to request that changes be made in the article
Software Requirements Specification	A document that completely describes all of the functions of a proposed system and the constraints under which it must operate. For example, this document
Stakeholder	Any person with an interest in the project who is not a developer
User	Reviewer or Author

1.4　References

IEEE Standard 830-1998 IEEE Recommended Practice for Software Requirements Specifications, IEEE Computer Society, 1998.

1.5　Overview of Document

The next chapter, Overall Description, of this document gives an overview of the product functionality. It describes the informal requirements and is used to establish a context for the technical requirements specification in next chapter.

The third chapter, Requirements Specification, of this document is written primarily for the developers, describing the product functionality details in technical terms.

Both sections of the document describe the same software product in its entirety, but are intended for different audiences and thus use different languages.

2. Overall Description

2.1　System Environment

Figure 2-4 illustrates the system environment.

The Web Publishing System has four active actors and one cooperating system.

The Author, Reader, or Reviewer accesses the Online Journal through the Internet. Any Author or Reviewer communication with the system is through E-mail. The Editor accesses the entire system directly. There is a link to the (existing) Historical Society Database.

《The division of the Web Publishing System into two component parts, the Online Journal and the Article Manager, is an example of using domain classes to make an explanation clearer.》

2.2　Functional Requirements Specification

This section outlines the use cases for each active actor separately. The reader, the

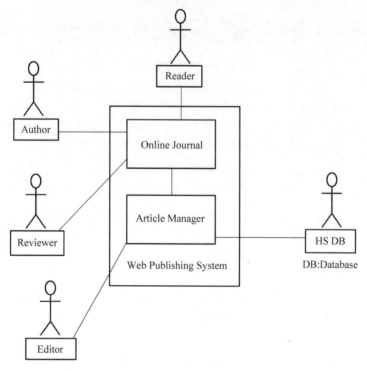

Figure 2-4 System Environment Diagram

author and the reviewer have only one use case apiece while the editor is the main actor in this system.

2.2.1 Reader Use Case

Use case: Search Article

Diagram

Figure 2-5 is the use case diagram for Search Article.

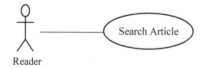

Figure 2-5 Use Case Diagram for Search Article

Brief Description

The Reader accesses the Online Journal website, searches for an article and downloads it to his/her machine.

Initial Step-By-Step Description

Before this use case can be initiated, the Reader has already had access to the Online Journal website.

(1) The Reader chooses to search by author name, category, or keyword.

(2) The system displays the choices to the Reader.

(3) The Reader selects the article desired.
(4) The system presents the abstract of the article to the Reader.
(5) The Reader chooses to download the article.
(6) The system provides the requested article.

Xref: Section 3.2.1, Search Article

The *Article Submission Process* state-transition diagram (Figure 2-6) summarizes the use cases listed below. An Author submits an article for consideration. The Editor enters it into the system and assigns and sends it to at least three reviewers. The Reviewers return their comments, which are used by the Editor to make a decision on the article. Either the article is accepted as written, declined, or the Author is asked to make some changes based on the reviews. If it is accepted, possibly after a revision, the Editor sends a copyright form to the Author. When that form is returned, the article is published to the Online Journal. The removal of a declined article from the system is not shown above.

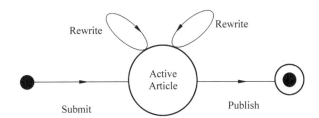

Figure 2-6 Article Submission Process State-transition Diagram

2.2.2 Author Use Case

In case of multiple authors, this term refers to the *principal author*, with whom all communication is made.

Use case: Submit Article

Diagram

Figure 2-7 is the use case diagram for Submit Article.

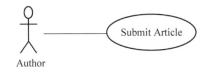

Figure 2-7 Use Case Diagram for Submit Article

Brief Description

The author either submits an original article or resubmits an edited article.

Initial Step-By-Step Description

Before this use case can be initiated, the Author has already connected to the Online Journal website.

(1) The Author chooses the *E-mail Editor* button.

(2) The System uses the *sendto* HTML tag to bring up the user's E-mail system.

(3) The Author fills in the Subject line and attaches the files as directed and E-mails them.

(4) The System generates and sends an E-mail acknowledgement.

Xref: Section 3.2.2, Communicate

2.2.3 Reviewer Use Case

Use case: Submit Review

Diagram

Figure 2-8 is the use case diagram for Submit Review.

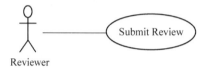

Figure 2-8　Use Case Diagram for Submit Review

Brief Description

The reviewer submits a review of an article.

Initial Step-By-Step Description

Before this use case can be initiated, the Reviewer has already connected to the Online Journal website.

(1) The Reviewer chooses the *E-mail Editor* button.

(2) The System uses the *sendto* HTML tag to bring up the user's E-mail system.

(3) The Reviewer fills in the Subject line, attaches the file as directed and E-mails it.

(4) The System generates and sends an E-mail acknowledgement.

Xref: Section 3.2.2, Communicate

2.2.4　Editor Use Cases

The Editor has the following sets of use cases (Figure 2-9):

1) Update Information Use Cases

(1) Use case: Update Author

Diagram

Figure 2-10 is the use case diagram for Update Author.

Brief Description

The Editor enters a new Author or updates information about a current Author.

Initial Step-By-Step Description

Before this use case can be initiated, the Editor has already had access to the main page of the Article Manager.

① The Editor selects to *Add/Update* Author.

② The system presents a choice of adding or updating.

Unit 2 Capturing the Requirements 需求获取

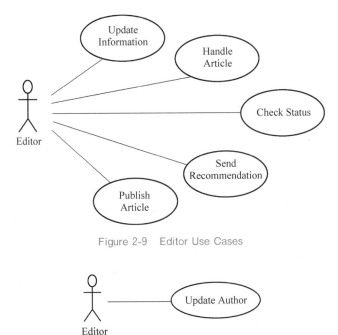

Figure 2-9 Editor Use Cases

Figure 2-10 Use Case Diagram for Update Author

③ The Editor chooses to add or update.

④ If the Editor is updating an Author, the system presents a list of authors to choose from and presents a grid filling in with the information; else the system presents a blank grid.

⑤ The Editor fills in the information and submits the form.

⑥ The system verifies the information and returns the Editor to the Article Manager main page.

Xref: Section 3.2.3, Add Author; Section 3.2.5 Update Person

(2) Use case: Update Reviewer

Diagram

Figure 2-11 is the use case diagram for Update Reviewer.

Figure 2-11 Use Case Diagram for Update Reviewer

Brief Description

The Editor enters a new Reviewer or updates information about a current Reviewer.

Initial Step-By-Step Description

Before this use case can be initiated, the Editor has already had access to the main page of the Article Manager.

053

① The Editor selects to *Add/Update* Reviewer.

② The system presents a choice of adding or updating.

③ The Editor chooses to add or update.

④ The system links to the Historical Society Database.

⑤ If the Editor is updating a Reviewer, the system presents a grid with the information about the Reviewer; else the system presents a list of members for the editor to select a Reviewer and presents a grid for the person selected.

⑥ The Editor fills in the information and submits the form.

⑦ The system verifies the information and returns the Editor to the Article Manager main page.

Xref: Section 3.2.4, Add Reviewer; Section 3.2.5, Update Person

(3) Use case: Update Article

Diagram

Figure 2-12 is the use case diagram for Update Article.

Figure 2-12 Use Case Diagram for Update Article

Brief Description

The Editor enters information about an existing article.

Initial Step-By-Step Description

Before this use case can be initiated, the Editor has already had access to the main page of the Article Manager.

① The Editor selects to *Update Article*.

② The system presents a list of active articles.

③ The system presents the information about the chosen article.

④ The Editor updates and submits the form.

⑤ The system verifies the information and returns the Editor to the Article Manager main page.

Xref: Section 3.2.6, Update Article Status

2) Handle Article Use Cases

(1) Use case: Receive Article

Diagram

Figure 2-13 is the use case diagram for Receive Article.

Brief Description

The Editor enters a new or revised article into the system.

Initial Step-By-Step Description

Before this use case can be initiated, the Editor has already had access to the main

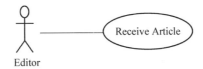

Figure 2-13   Use Case Diagram for Receive Article

page of the Article Manager and had a file containing the article available.

① The Editor selects to *Receive Article*.

② The system presents a choice of entering a new article or updating an existing article.

③ The Editor chooses to add or update.

④ If the Editor is updating an article, the system presents a list of articles to choose from and presents a grid for filling with the information; else the system presents a blank grid.

⑤ The Editor fills in the information and submits the form.

⑥ The system verifies the information and returns the Editor to the Article Manager main page.

Xref: Section 3.2.7, Enter Communication

(2) Use case: Assign Reviewer

This use case extends the *Update Article* use case.

Diagram

Figure 2-14 is the use case diagram for Assign Reviewer.

Figure 2-14   Use Case Diagram for Assign Reviewer

Brief Description

The Editor assigns one or more reviewers to an article.

Initial Step-By-Step Description

Before this use case can be initiated, the Editor has already had access to the article using the *Update Article* use case.

① The Editor selects to *Assign Reviewer*.

② The system presents a list of Reviewers with their status (see data description in section 3.3 below).

③ The Editor selects a Reviewer.

④ The system verifies that the person is still an active member using the Historical Society Database.

⑤ The Editor repeats steps 3 and 4 until sufficient reviewers are assigned.

⑥ The system E-mails the Reviewers, attaching the article and requesting reviews.

⑦ The system returns the Editor to the *Update Article* use case.

Xref: Section 3.2.8, Assign Reviewer

(3) Use case: Receive Review

This use case extends the *Update Article* use case.

Diagram

Figure 2-15 is the use case diagram for Receive Review.

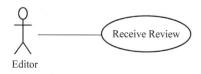

Figure 2-15 Use Case Diagram for Receive Review

Brief Description

The Editor enters a review into the system.

Initial Step-By-Step Description

Before this use case can be initiated, the Editor has already had access to the article using the *Update Article* use case.

① The Editor selects to *Receive Review*.

② The system presents a form for filling with the information.

③ The Editor fills in the information and submits the form.

④ The system verifies the information and returns the Editor to the Article Manager main page.

Xref: Section 3.2.7, Enter Communication

3) Check Status Use Case

Use case: Check Status

Diagram

Figure 2-16 is the use case diagram for Check Status.

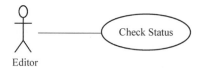

Figure 2-16 Use Case Diagram for Check Status

Brief Description

The Editor checks the status of all active articles.

Initial Step-By-Step Description

Before this use case can be initiated, the Editor has already had access to the main page of the Article Manager.

① The Editor selects to *Check Status*.

② The system returns a scrollable list of all active articles with their status (see data description in section 3.3 below).

③ The system returns the Editor to the Article Manager main page.

Xref: Section 3.2.9, Check Status

4) Send Recommendation Use Cases

(1) Use case: Send Response

This use case extends the *Update Article* use case.

Diagram:

Figure 2-17 is the use case diagram for Send Response.

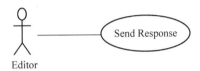

Figure 2-17   Use Case Diagram for Send Response

Brief Description

The Editor sends a response to an Author.

Initial Step-By-Step Description

Before this use case can be initiated, the Editor has already had access to the article using the *Update Article* use case.

① The Editor selects to *Send Response*.

② The system calls the E-mail system and puts the Author's E-mail address in the Recipient line and the name of the article on the Subject line.

③ The Editor fills in the E-mail text and sends the message.

④ The system returns the Editor to the Article Manager main page.

Xref: Section 3.2.10, Send Communication

(2) Use case: Send Copyright

This use case extends the *Update Article* use case.

Diagram

Figure 2-18 is the use case diagram for Send Copyright.

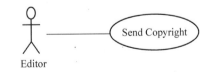

Figure 2-18   Use Case Diagram for Send Copyright

Brief Description

The Editor sends a copyright form to an Author.

Initial Step-By-Step Description

Before this use case can be initiated, the Editor has already had access to the article using the Update Article use case.

① The Editor selects to *Send Copyright*.

② The system calls the E-mail system, puts the Author's E-mail address and the name of the article in the Recipient line and the Subject line respectively, and attaches the copyright form.

③ The Editor fills in the E-mail text and sends the message.

④ The system returns the Editor to the Article Manager main page.

Xref: Section 3.2.10, Send Communication

(3) Use case: Remove Article

This use case extends the *Update Article* use case.

Diagram

Figure 2-19 is the use case diagram for Remove Article.

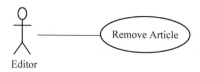

Figure 2-19 Use Case Diagram for Remove Article

Brief Description

The Editor removes an article from the active category.

Initial Step-By-Step Description

Before this use case can be initiated, the Editor has already had access to the article using the *Update Article* use case.

① The Editor selects to remove an article from the active database.

② The system provides a list of articles with the status of each.

③ The Editor selects an article for removal.

④ The system removes the article from the active article database and returns the Editor to the Article Manager main page.

Xref: Section 3.2.12, Remove Article

5) Publish Article Use Case

Use case: Publish Article

This use case extends the *Update Article* use case.

Diagram

Figure 2-20 is the use case diagram for Publish Article.

Brief Description

The Editor transfers an accepted article to the Online Journal.

Unit 2　Capturing the Requirements　需求获取

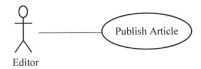

Figure 2-20　Use Case Diagram for Publish Article

Initial Step-By-Step Description

Before this use case can be initiated, the Editor has already had access to the article using the *Update Article* use case.

① The Editor selects to *Publish Article*.

② The system transfers the article to the Online Journal and updates the search information there.

③ The system removes the article from the active article database and returns the Editor to the Article Manager home page.

Xref: Section 3.2.11, Publish Article

《Since three of the actors only have one use case each, the summary diagram only involves the Editor. We should adapt the rules to the needs of the document rather than adapt the document to fit the rules.》

2.3　User Characteristics

The Reader is expected to be Internet literate and be able to use a search engine. The main screen of the Online Journal website will have the search function and a link to "Author/Reviewer Information."

The Author and Reviewer are expected to be Internet literate and to be able to use E-mail with attachments.

The Editor is expected to be Windows literate and to be able to use button, pull-down menus, and similar tools.

The detailed look of these pages is discussed in section 3.2 below.

2.4　Non-Functional Requirements

The Online Journal will be on a server with high speed Internet capability. The physical machine to be used will be determined by the Historical Society. The software developed here assumes the use of a tool such as socket for transfer among different servers and uses Visual Studio which supports SQL Server's links to implement the import and export of databases. The speed of the Reader's connection will depend on the hardware used rather than characteristics of this system.

The Article Manager will run on the editor's PC and will contain an SQL Server database. SQL Server is already installed on this computer and in a Windows operating system.

3. Requirements Specification

3.1 External Interface Requirements

The only link to an external system is the link to the Historical Society (HS) Database to verify the membership of a Reviewer. The Editor believes that a society member is much more likely to be an effective reviewer and has imposed a membership requirement for a Reviewer. The HS Database fields of interest to the Web Publishing Systems are member name, membership (ID) number, and E-mail address (an optional field for the HS Database).

The *Assign Reviewer* use case sends the Reviewer ID to the HS Database and a Boolean is returned denoting the membership status. The *Update Reviewer* use case requests a list of member names, membership numbers and (optional) E-mail addresses when adding a new Reviewer. It returns a Boolean for membership status when updating a Reviewer.

3.2 Functional Requirements

The Logical Structure of the Data is contained in section 3.3.1.

3.2.1 Search Article

Table 2-2 illustrates the Search Article use case.

Table 2-2 Search Article Use Case

Use Case Name	Search Article
XRef	Section 2.2.1, Search Article SDS, Section 8.1
Trigger	The Reader accesses the Online Journal website
Pre-condition	The Web is displayed with forms for searching
Basic Path	1. The Reader chooses how to search the website. The choices are by Author, by Category, and by Keyword. 2. If the search is by Author, the system creates and presents an alphabetical list of all authors in the database. In the case of an article with multiple authors, each is contained in the list. 3. The Reader selects an author. 4. The system creates and presents a list of all articles by that author in the database. 5. The Reader selects an article. 6. The system displays the Abstract for the article. 7. The Reader selects to download the article or to return to the article list or to the previous list

continued

Use Case Name	Search Article
Alternative Paths	In step 2, if the Reader selects to search by Category, the system creates and presents a list of all categories in the database. 3. The Reader selects a category. 4. The system creates and presents a list of all articles in that category in the database. Return to step 5. In step 2, if the Reader selects to search by Keyword, the system presents a dialog box to enter the keyword or phrase. 3. The Reader enters a keyword or phrase. 4. The system searches the Abstracts for all articles with that keyword or phrase and creates and presents a list of all such articles in the database. Return to step 5
Post-condition	The selected article is downloaded to the client machine
Exception Paths	The Reader may abandon the search at any time
Other	The categories list is generated from the information provided when article are published and not predefined in the Online Journal database

3.2.2 Communicate

Table 2-3 illustrates the Communicate use case.

Table 2-3 Communicate Use Case

Use Case Name	Communicate
XRef	Section 2.2.2, Submit Article; Section 2.2.3, Submit Review SDS, Section 8.2
Trigger	The user selects a mailto link
Pre-condition	The user is on the Communicate page linked from the Online Journal main page
Basic Path	This use case uses the mailto HTML tag. This invokes the client E-mail facility
Alternative Paths	If the user prefers to use his or her own E-mail directly, sufficient information will be contained on the Web page to do so
Post-condition	The message is sent
Exception Paths	The attempt may be abandoned at any time
Other	None

3.2.3 Add Author

Table 2-4 illustrates the Add Author use case.

Table 2-4 Add Author Use Case

Use Case Name	Add Author
XRef	Section 2.2.4, Update Author SDS, Section 8.3
Trigger	The Editor selects to add a new author to the database

continued

Use Case Name	Add Author
Pre-condition	The Editor has had access to the Article Manager main screen
Basic Path	The system presents a blank form to enter the author information The Editor enters the information and submits the form The system checks that the name and E-mail address fields are not blank and updates the database
Alternative Paths	If in step 2, either field is blank, the Editor is instructed to add an entry. No validation for correctness is made
Post-condition	The Author has been added to the database
Exception Paths	The Editor may abandon the operation at any time
Other	The Author information includes the name, mailing address and E-mail address

3.2.4 Add Reviewer

Table 2-5 illustrates the Add Reviewer use case.

Table 2-5 Add Reviewer Use Case

Use Case Name	Add Reviewer
XRef	Section 2.2.4, Update Reviewer SDS, Section 8.4
Trigger	The Editor selects to add a new reviewer to the database
Pre-condition	The Editor has had access to the Article Manager main screen
Basic Path	1. The system accesses the Historical Society (HS) database and presents an alphabetical list of the society members. 2. The Editor selects a person. 3. The system transfers the member information from the HS database to the Article Manager (AM) database. If there is no E-mail address in the HS database, the Editor is prompted for an entry in that field. 4. The information is entered into the AM database
Alternative Paths	In step 3, if there is no entry for the E-mail address in the HS database or on this form, the Editor will be re-prompted for an entry. No validation for correctness is made
Post-condition	The Reviewer has been added to the database
Exception Paths	The Editor may abandon the operation at any time
Other	The Reviewer information includes name, membership number, mailing address, categories of interest, and E-mail address

3.2.5 Update Person

Table 2-6 illustrates the Update Person use case.

Table 2-6 Update Person Use Case

Use Case Name	Update Person
XRef	Section 2.2.4 Update Author; Section 2.2.4 Update Reviewer SDS, Section 8.5
Trigger	The Editor selects to update an author or reviewer and the person is already in the database
Pre-condition	The Editor has had access to the Article Manager main screen
Basic Path	1. The Editor selects Author or Reviewer 2. The system creates and presents an alphabetical list of people in the category 3. The Editor selects a person to update 4. The system presents the database information in the form for modification 5. The Editor updates the information and submits the form 6. The system checks that required fields are not blank
Alternative Paths	In step 5, if any required field is blank, the Editor is instructed to add an entry. No validation for correctness is made
Post-condition	The database has been updated
Exception Paths	If the person is not already in the database, the use case is abandoned. In addition, the Editor may abandon the operation at any time
Other	This use case is not used when one of the other use cases is more appropriate, such as to add an article or a reviewer for an article

3.2.6 Update Article Status

Table 2-7 illustrates the Update Article Status use case.

Table 2-7 Update Article Status Use Case

Use Case Name	Update Article Status
XRef	Section 2.2.4, Update Article SDS, Section 8.6
Trigger	The Editor selects to update the status of an article in the database
Pre-condition	The Editor has had access to the Article Manager main screen and the article is already in the database
Basic Path	1. The system creates and presents an alphabetical list of all active articles 2. The Editor selects the article to update 3. The system presents the information about the article in the form format 4. The Editor updates the information and resubmits the form
Alternative Paths	In step 4, the use case *Enter Communication* may be invoked
Post-condition	The database has been updated
Exception Paths	If the article is not already in the database, the use case is abandoned. In addition, the Editor may abandon the operation at any time
Other	This use case can be used to add categories for an article, to correct typographical errors, or to remove a reviewer who has missed a deadline for returning a review. It may also be used to allow access to the named use case to enter an updated article or a review for an article

3.2.7 Enter Communication

Table 2-8 illustrates the Enter Communication use case.

Table 2-8 Enter Communication Use Case

Use Case Name	Enter Communication
XRef	Section 2.2.4, Receive Article; Section 2.2.4, Receive Review SDS, Section 8.7
Trigger	The Editor selects to add a document to the system
Pre-condition	The Editor has had access to the Article Manager main screen and has the file of the item to be entered available
Basic Path	1. The Editor selects the article using the 3.2.6, Update Article Status use case 2. The Editor attaches the file to the form presented and updates the respective information about the article 3. When the Editor updates the article status to indicate that a review is returned, the respective entry in the Reviewer table is updated
Alternative Paths	None
Post-condition	The article entry is updated in the database
Exception Paths	The Editor may abandon the operation at any time
Other	This use case extends 3.2.6, Update Article Status

3.2.8 Assign Reviewer

Table 2-9 illustrates the Assign Reviewer use case.

Table 2-9 Assign Reviewer Use Case

Use Case Name	Assign Reviewer
XRef	Section 2.2.4, Assign Reviewer SDS, Section 8.8
Trigger	The Editor selects to assign a reviewer to an article
Pre-condition	The Editor has had access to the Article Manager main screen and the article is already in the database
Basic Path	1. The Editor selects the article using the 3.2.6, Update Article Status use case 2. The system presents an alphabetical list of reviewers with their information 3. The Editor selects a reviewer for the article 4. The system updates the article database entry and E-mails the reviewer with the standard message and attaches the text of the article without author information 5. The Editor has the option of repeating this use case from step 2
Alternative Paths	None
Post-condition	At least one reviewer has been added to the article information and the appropriate communication has been sent
Exception Paths	The Editor may abandon the operation at any time

Unit 2 Capturing the Requirements 需求获取

continued

Use Case Name	Assign Reviewer
Other	This use case extends 3.2.6, Update Article Status. The Editor, prior to implementation of this use case, will provide the message text

3.2.9 Check Status

Table 2-10 illustrates the Check Status use case.

Table 2-10 Check Status Use Case

Use Case Name	Check Status
XRef	Section 2.2.4, Check Status SDS, Section 8.9
Trigger	The Editor has selected to check status of all active articles
Pre-condition	The Editor has had access to the Article Manager main screen
Basic Path	The system creates and presents a list of all active articles organized by their status. The Editor may request to see the full information about an article
Alternative Paths	None
Post-condition	The requested information has been displayed
Exception Paths	The Editor may abandon the operation at any time
Other	The editor may provide an enhanced list of status later. At present, the following categories must be provided: 1. Received but no further action taken 2. Reviewers have been assigned but not all reviews are returned (including dates that reviewers were assigned and order by this criterion) 3. Reviews returned but no further action taken 4. Recommendations for revision sent to Author but no response as of yet 5. Author has revised article but no action has been taken 6. Article has been accepted and copyright form has been sent 7. Copyright form has been returned but article is not yet published A published article is automatically removed from the active article list

3.2.10 Send Communication

Table 2-11 illustrates the Send Communication use case.

Table 2-11 Send Communication Use Case

Use Case Name	Send Communication
XRef	Section 2.2.4, Send Response; Section 2.2.4, Send Copyright SDS, Section 8.10
Trigger	The Editor selects to send a communication to an author
Pre-condition	The Editor has had access to the Article Manager main screen

continued

Use Case Name	Send Communication
Basic Path	1. The system presents an alphabetical list of authors 2. The Editor selects an author 3. The system invokes the Editor's E-mail system entering the author's E-mail address into the To: entry 4. The Editor uses the E-mail facility
Alternative Paths	None
Post-condition	The communication has been sent
Exception Paths	The Editor may abandon the operation at any time
Other	The standard copyright form will be available in the Editor's directory for attaching to the E-mail message, if desired

3.2.11 Publish Article

Table 2-12 illustrates the Publish Article use case.

Table 2-12 Publish Article Use Case

Use Case Name	Publish Article
XRef	Section 2.2.4, Publish Article SDS, Section 8.11
Trigger	The Editor selects to transfer an approved article to the Online Journal
Pre-condition	The Editor has had access to the Article Manager main screen
Basic Path	1. The system creates and presents an alphabetical list of the active articles that are flagged as having their copyright form returned. 2. The Editor selects an article to publish. 3. The system accesses the Online Database and transfers the article and its accompanying information to the Online Journal database. 4. The article is removed from the active article database
Alternative Paths	None
Post-condition	The article is properly transferred
Exception Paths	The Editor may abandon the operation at any time
Other	Find out from the Editor to see if the article information should be archived somewhere

3.2.12 Remove Article

Table 2-13 illustrates the Remove Article use case.

Table 2-13 Remove Article Use Case

Use Case Name	Remove Article
XRef	Section 2.2.4, Remove Article SDS, Section 8.12

continued

Use Case Name	Remove Article
Trigger	The Editor selects to remove an article from the active article database
Pre-condition	The Editor has had access to the Article Manager main screen
Basic Path	1. The system provides an alphabetized list of all active articles 2. The Editor selects an article 3. The system displays the information about the article and requires that the Editor confirm the deletion 4. The Editor confirms the deletion
Alternative Paths	None
Post-condition	The article is removed from the database
Exception Paths	The Editor may abandon the operation at any time
Other	Find out from the Editor to see if the article and its information should be archived somewhere

3.3 Detailed Non-Functional Requirements

3.3.1 Logical Structure of the Data

The logical structure of the data to be stored in the internal Article Manager database is given in Figure 2-21.

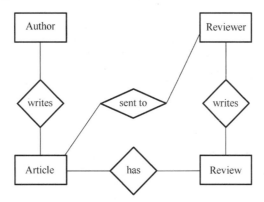

Figure 2-21 Logical Structure of the Article Manager Data

The data descriptions of each of these data entities are as follows:

(1) Table 2-14 illustrates the Author data entity.

Table 2-14 Author Data Entity

Data Item	Type	Description	Comment
Name	Varchar (80)	Name of principle author	
E-mail Address	Varchar (100)	Internet address	
Article	Cursor	Article entity	May be several

(2) Table 2-15 illustrates the Reviewer data entity.

Table 2-15 Reviewer Data Entity

Data Item	Type	Description	Comment
Name	Varchar (80)	Name of principle author	
ID	Integer	ID number of Historical Society member	Used as key in Historical Society Database
E-mail Address	Varchar (100)	Internet address	
Article	Cursor	Article entity	May be several
Num Review	Integer	Review count	Number of not returned reviews (updated every time when a review status is changed)
History	Text	Comments on past performance	
Specialty	Varchar (20)	Area of expertise	May be several

(3) Table 2-16 illustrates the Review data entity.

Table 2-16 Review Data Entity

Data Item	Type	Description	Comment
Article	Cursor	Article entity	
Reviewer	Cursor	Reviewer entity	Single reviewer
Date Sent	Date	Date sent to reviewer	
Returned	Date	Date returned; null if not returned	
Contents	Text	Text of review	

(4) Table 2-17 illustrates the Article data entity.

Table 2-17 Article Data Entity

Data Item	Type	Description	Comment
Name	Varchar (80)	Name of Article	
Author	Cursor	Author entity	Name of principle author
Other Authors	Varchar	Other authors is any; else null	Not a pointer to an Author entity
Reviewer	Cursor	Reviewer entity	Will be several
Review	Cursor	Review entity	Set up when reviewer is set up
Contents	Text	Body of article	Contains Abstract as first paragraph.
Abstract	Text	Abstract of article	
Category	Varchar (20)	Area of content	May be several
Accepted	Bit	Article has been accepted for publication	Needs Copyright form returned

continued

Data Item	Type	Description	Comment
Copyright	Bit	Copyright form has been returned	Not relevant unless Accepted is True.
Published	Bit	Sent to Online Journal	Not relevant unless Accepted is True. Article is no longer active and does not appear in status checks

The Logical Structure of the data to be stored in the Online Journal database on the server is shown in Table 2-18.

Table 2-18　Published Article Data Entity

Data Item	Type	Description	Comment
Name	Varchar (80)	Name of Article	
Author	Varchar (50)	Name of one Author	May be several
Abstract	Text	Abstract of article	Used for keyword search
Content	Text	Body of article	
Category	Text	Area of content	May be several

3.3.2　Security

The server on which the Online Journal resides will have its own security to prevent unauthorized *write/delete* access. There is no restriction on read access. The use of E-mail by an Author or Reviewer is on the client systems and thus is external to the system.

The PC on which the Article Manager resides will have its own security. Only the Editor will have physical access to the machine and the program on it. There is no special protection built into this system other than to provide the editor with write access to the Online Journal to publish an article.

 Exercises

1. How many actors are there in this Web Publishing System?
2. What does the use case Editor do?
3. How many data items are there in the data entity Articles?

Unit 3

Planning the Project

项目策划

Unit 3　Planning the Project　项目策划

Part 1

Listening & Speaking

Unit 3

Dialogue: Software Project Planning

Jason: So much work to do.

Kevin: I think we need a formal project plan as our guideline.

Sharon: Yes, the first one is the time, which is one of the most important factors for a project. We need a schedule, especially, the deadline of our project.

Kevin: We have 40 days in total; the requirements acquirement has taken us 5 days already, so there are 35 days left.

Jason: Oh, it sounds really urgent.

Sharon: It seems that we should begin to program as soon as possible, right?

Jason: Although coding is a very central part of a software project, the most important thing is that we must first establish a proper schedule to control our progress and assure our **deployment** on time, I think.

Kevin: Yes. In the requirements phase, we will spend another 3 days to depict, analyze and model the requirements. After that, we will spend 3 days to complete the **architectural design** and it will take 5 days to accomplish the **detailed design**. Because of the effort applied to software design, code should follow with relatively little difficulty, and can be done within one week I think. Testing and **subsequent debugging** can **account for** about 10 days of software development effort.

071

Sharon: Maybe we can draw our schedule using a **Gantt chart**. It is a visual and effective tool for a project plan.

Kevin: Good idea.

Jason: Actually, testing should not be seen as an activity which starts only after the coding phase is complete, with the limited purposes of detecting failures. [1] Indeed, planning for testing should start with the early stages of the requirement process, and test plans and procedures must be systematically and continuously developed, and possibly refined, as development proceeds. During coding, we can perform the unit testing at the same time. It will save much time and obtain better testing effects I think. Finally, we can perform a validation test by working with the customer to find out if the software developed is valid for the customer and make sure that the customer is getting what they asked for.

[1] Replace with: Software design is an **indispensable** process which has a significant impact on the next processes in the entire development.

Sharon: By the way, we need three computers and must install the software which the customer requires with a uniform version as developing tools. That is Microsoft Visual Studio 2017 for a development platform and Microsoft SQL server 2016 as a database management system.

Jason: We need a network as well.

Kevin: Ok, I will prepare the development environment for us as soon as possible. Then, we must assign some management responsibilities to everyone. Sharon, you are responsible for document management [2]. Jason, you take charge of change management. And I will be in charge of the Software Quality Assurance. Ok?

[2] Replace with:
1. database management
2. **configuration** management
3. risk management

Jason & Sharon: No problem.

Kevin: Ok, I have **noted** everything which we **referred to** just now, and will complete a project plan within two days.

 Exercises

Work in a group, and make up a similar conversation by replacing the statements with other expressions on the right side.

 Words

deployment[di'plɔimənt] *n.* 调度,部署
subsequent['sʌbsikw(ə)nt]*adj.* 随后的,继……之后的
debug[ˌdiː'bʌg] *v.* 调试

indispensable[ˌindi'spensəbl] *adj.* 不可缺少的,必需的
configuration[kənˌfigə'reiʃn] *n.* 配置
note[nəut] *v.* 记录

 Phrases

architectural design　体系结构设计,概要设计
detailed design　详细设计
account for　（在数量、比例方面）占
Gantt chart　甘特图
refer to　提到

Listening Comprehension: Software Project Planning

Listen to the article and answer the following 3 questions based on it. After you hear a question, there will be a break of 15 seconds. During the break, you will decide which one is the best answer among the four choices marked (A), (B), (C) and (D).

Questions

1. How many kinds of planning philosophies are mentioned in the article?
   (A) Two　　　　　　　　(B) Three
   (C) Four　　　　　　　　(D) Five

2. How many questions are stated in Boehm's principle for leading project planning?
   (A) Three　　　　　　　(B) Five
   (C) Six　　　　　　　　(D) Seven

073

3. Which point of view is the most accordant with the idea of this article on the project planning?

（A）Planning every activity in the project as detailed as possible for their foreseeable ability

（B）Carrying out as early as possible regardless of planning, because even the best planning can be obviated by uncontrolled change as the work proceeds

（C）Adjusting different levels of details for activities according to their different locations in the project timeline

（D）Making a perfect plan which can evade changes that may come about during the work

 Words

philosophy[fəˈlɔsəfi] n. 方法
minimalist [ˈminiməlist] n. 最低限要求者
argue[ˈɑːgjuː] v. 主张, 认为, 表明, 论证
obviate[ˈɔbvieit] v. 排除, 避免
traditionalist[trəˈdiʃənəlist] n. 传统主义者, 墨守成规者
agilist [ˈædʒailist] n. 敏捷主义者, 机敏者

fruitless[ˈfruːtləs] adj. 徒劳的, 无用的
recipe[ˈresəpi] n. 诀窍
chaos[ˈkeiɔs] n. 混乱
paper[ˈpeipə(r)] n. 论文, 文章
state[steit] v. 陈述, 说明
milestone[ˈmailstəun] n. 里程碑
resultant[riˈzʌltənt] adj. 因此而产生的
managerially[ˌmænəˈdʒiəriəli] adv. 管理地

 Phrases

road map   线路图, (一步一步的)详尽计划
in moderation   适中地
scale down   按比例减少, 按比例缩小

 Abbreviations

W5HH   Why, What, When, Who, Where, How, How much 为什么, 什么, 什么时候, 谁, 在哪里, 如何, 有多少

Dictation: Four Variables in Projects

This article will be played three times. Listen carefully, and fill in the numbered spaces with the appropriate words you have heard.

Unit 3 Planning the Project 项目策划

A project is a carefully defined set of ___1___ that use resources to achieve ___2___ goals and objectives. It is a *finite endeavor* ___3___ specific start and completion dates as well as a managed ___4___ having a range of ___5___, budget and organizational constraints.

It is usually considered that there are four ___6___ variables we will ___7___ in software projects—cost, time, scope, and quality. These four variables affect a project together by ___8___ with each other: increased scope typically means more time and ___9___, a tight time *stress* could mean stronger finance support and *suffered* ___10___, too little money couldn't solve the customer's business problem ___11___ the scope, and a higher quality might deliver longer developing ___12___ and more cost.

It seems that there is not a simple relationship between them. For example, you can't just get software faster by spending more money. As the *saying* ___13___, "Nine women cannot make a ___14___ in one month."

In his book *eXtreme Programming* [1] Explained: *Embrace* Change, Kent Beck says that the solution is to make the four variables ___15___. If everyone—programmers, customers, and managers—can see all four variables, they can *consciously* ___16___ which variables to control. If they don't like the result ___17___ for fourth variable, they can change the ___18___, or they can ___19___ a different three variables to control for ___20___ *ultimately* the project objectives.

Words

finite ['fainait] *adj.* 有限的,限定的	embrace [im'breis] *v.* 掌握,接受,拥抱
endeavor [in'devə] *n.* 努力,尽力	consciously ['kɔnʃəsli] *adv.* 有意识地
stress [stres] *n.* 压力,紧迫	ultimately ['ʌltimətli] *adv.* 最后,终于,
suffer ['sʌfə(r)] *v.* 受损失,受害	根本
saying ['seiiŋ] *n.* 谚语,格言	

Phrases

eXtreme Programming 极限编程

Notes

[1] 极限编程 (eXtreme Programming, XP) 是一种软件工程方法学,是敏捷软件开发中最富有成效的几种方法学之一。如同其他敏捷方法学,极限编程和传统方法学的本质不

同在于它更强调可适应性以及面临的困难,适用于小团队开发。

Part 2

Reading & Translating

Section A: Software Project Plan

A proper project plan is an important **ingredient** for a successful project. Without proper planning, a software development project is unlikely to succeed. Good planning can be done after the requirements and architecture for the project are available. The important planning activities are: process planning, effort estimation, scheduling and **staffing** planning, quality planning, configuration management planning, project monitoring planning, and risk management.

Process planning generally involves selecting a proper process model and **tailoring** it to **suit** the project needs. In effort estimation, overall effort requirement for the project and the **breakup** of the effort for different phases is estimated. In a top-down approach, total effort is first estimated, frequently from the estimate of the size, and then effort for different phases or tasks is determined. In a bottom-up approach, the main tasks in the project are **identified**, and effort for them is estimated first. From the effort estimates of the tasks, the overall estimate is obtained.

The overall schedule and the major milestones of a project depend on the effort estimate and the staffing level in the project and simple models can be used to get a rough estimate of schedule from effort. Often, an overall schedule is determined using a model, and then adjusted to meet the project needs and constraints. The detailed schedule is one in which the tasks are broken into smaller, schedulable tasks, and then assigned to specific team members, while preserving the overall schedule and effort estimates. [1] The detailed schedule is the most **live** document of project planning as it lists the tasks that have to be done; any changes in the project plan must be reflected suitably in the detailed schedule (Figure 3-1).

Quality plans are important for ensuring that the final product is of high quality. The project quality plan identifies all the **V&V** activities that have to be performed at different stages in the development, and how they are to be performed.

The goal of configuration management is to control the changes that take place during the project. The configuration management plan identifies the configuration **items** which will be controlled, and specifies the procedures to accomplish this and how access is to be controlled.

Unit 3 Planning the Project 项目策划

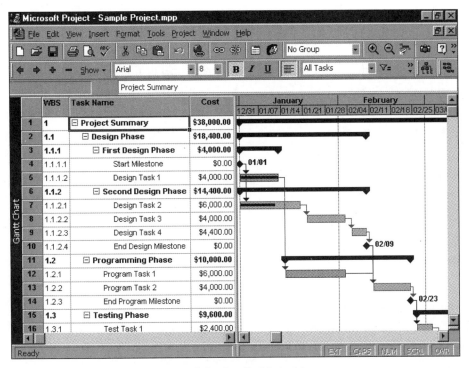

Figure 3-1 Detailed Schedule

Risks are those events which may or may not occur, but if they do occur, they have a negative impact on the project. To meet project goals even under the presence of risks requires proper risk management. Risk management requires that risks be identified, analyzed, and prioritized. Then risk mitigation plans are made and performed to minimize the effect of the highest priority risks.

For a plan to be successfully implemented it is essential that the project be monitored carefully. Activity level monitoring, status reports, and milestone analysis are the mechanisms that are often used. For analysis and reports, the actual effort, schedule, defects, and size should be measured. With these measurements, it is possible to monitor the performance of a project with respect to its plan. And based on this monitoring, actions can be taken to correct the course of execution, if the need arises.

Overall, project planning lays out the path the project should follow in order to achieve the project objectives. It specifies all the tasks that the project members should perform, and specifies who will do what, in how much time, and when in order to execute this plan. With a detailed plan, what remains to be done is to execute the plan, which is done through the rest of the project. Of course, plans never remain unchanged, as things do not always work as planned. With proper monitoring in place, these situations can be identified and plans changed accordingly. Basic project planning principles and techniques can be used for plan modification also.

Words

ingredient[inˈɡriːdiənt] n. 因素,成分
staff[stɑːf] v. 配置职员
tailor[ˈteilə(r)] v. 剪裁,适应
suit[suːt,sjuːt] v. 适合,合乎……的要求
breakup[ˈbreikʌp] n. 分解
identify[aiˈdentifai] v. 确认,说明身份
live[laiv] adj. 生动的
item[ˈaitəm] n. 项

prioritize[praiˈɔrətaiz] v. 把……区分优先次序
mitigation[ˌmitiˈɡeiʃn] n. 缓解,减轻
mechanism[ˈmekənizəm] n. 机制
defect[diːˈfekt,diˈfekt] n. 缺陷
course[kɔːs] n. 过程,进程
arise[əˈraiz] v. 出现,产生,上升

Phrases

with respect to   就……而论,关于
lay out   划定(路线),布置,安排
in place   适当,就位,(法律、政策、行政体系等)正在运作

Abbreviations

V&V   Verification and Validation   验证和确认

Notes

[1] Original: The detailed schedule is one in which the tasks are broken into smaller, schedulable tasks, and then assigned to specific team members, while preserving the overall schedule and effort estimates.

Translation: 详细进度是指在保持总体进度和工作量估算的条件下,将任务分解为更小的、可安排的任务,然后将其分派给特定的团队成员。

Exercises

Ⅰ. Read the following statements carefully, and decide whether they are true (T) or false (F) according to the text.

____ 1. There are seven important activities in the project planning.

____ 2. Generally, there are two main steps involved in process planning, that is selecting a proper process model and tailoring the project to suit the model chosen.

____ 3. In software engineering, the term "activity" is in the level with more details

Unit 3　Planning the Project　项目策划

than the term "task" is.

_____ 4. Both the effort estimate and the staffing level in a project are the bases of the overall schedule and the major milestones of the project.

_____ 5. Status reports is one of the mechanisms that are often used in project monitoring.

Ⅱ. Choose the best answer to each of the following questions according to the text.

1. Which statement is right about the project planning?
   (A) Planning is an important ingredient for a successful project, so it must be done at the very beginning, before any other activities in the project.
   (B) Project planning arranges the path the project should follow in order to achieve the project objectives.
   (C) Establishing a good project plan is the most essential for implementing the plan successfully.
   (D) Because of the potential risks brought by changes, a good project plan always avoids changes, so that things always work as planned.

2. What does the word "live" means in the context of "The detailed schedule is the most live document of project planning" in the third paragraph?
   (A) Involving team members who are physically present, because the tasks must be assigned to specific team members in the detailed schedule.
   (B) Guidable, because the detailed schedule lists the tasks that have to be done, as the guidance of the daily work.
   (C) Operative, because the detailed schedule must consider any changes in the project plan suitably.
   (D) Not specific, because the detailed schedule does not have to be made due to the potential frequent changes.

3. Which statement is wrong about the following different activities in the project plan?
   (A) Models that can be used in process planning and schedule planning are different.
   (B) Configuration management is used to control the changes that occur during the project.
   (C) Risk mitigation plans are made and performed as the subsequence of the risk management plan.
   (D) Measurement is a most important mechanism for monitoring the performance of a project with respect to its plan.

Ⅲ. Fill in the numbered spaces with the words or phrases chosen from the box. Change the forms where necessary.

> schedule   pick   check   ahead   chart
> through   no   integrate   assemble   take

Scheduling People

Your tasks aren't going to complete themselves. That's why you have ___1___ a team, but if that team isn't scheduled the way you have carefully ___2___ your task list, then you're not managing your project.

Over the course of a project's lifecycle team members are going to ___3___ off for holidays, personal days or vacation. If you're not prepared for these times, and have scheduled other team members to ___4___ up the slack in their absence, your schedule will suffer.

___5___ your calendar into a project management software is a simple way to stay on top of your resources. There's ___6___ reason to use a standalone calendar that sends you to another application every time you need to ___7___ on a team member's availability.

Another way to stay on top of your scheduling is by integrating your task scheduling view on the Gantt ___8___ with resource and workload scheduling features. You can schedule your team's workload ___9___ color-coding, so you know at-a-glance who is behind, ___10___ or on schedule with their tasks.

Ⅳ. Translate the following passages into Chinese.
Inaccuracy in Effort Estimation

Naturally, the importance and difficulty of project planning and especially effort estimation varies by teams, companies and projects. Compared to Agile planning, where teams usually only plan the next short sprint (and are even able to use feedback from previous sprints or releases), planning an entire Waterfall SDLC process can be a challenging task.

Bias from price-sensitive clients, relying on subjective judgement (however expert it may be) or estimates from a single source as opposed to several sources, and mainly the lack of historical data are all factors that negatively impact estimation accuracy. While some of these risks can be mitigated by organizational changes and revised management methods, the latter also requires the use of suitable tools. Collecting and organizing historical data, and analyzing it to improve the accuracy of estimates for future projects can only be efficiently performed by using software tools that support these processes.

Section B: AI-powered Project Management in the Near Future

Around a decade ago, most of us were skeptical about the application of Artificial Intelligence (AI) **just about** everywhere in business and life. We thought that AI could only be used for doing repetitive tasks. We created **scripts** and programmed computers to do simple, **routine** activities. But with big data, the growing processing power and advanced algorithms, we now know that AI systems can be designed to perform complex tasks and that deep learning can make a computer think and act like a human (Figure 3-2).

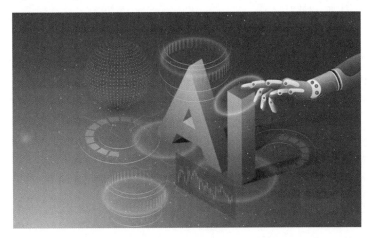

Figure 3-2 AI System

There are at least 3 ways AI can improve project management in the near future:

1. Less projects fail thanks to predictive analytics

Throughout its life-cycle, any project encounters a series of risks and uncertainties that can cause a failure. The team needs to regularly assess and respond to all the risks they identify based on their experience, knowledge and available tools. The challenge with the ongoing risk evaluation can turn into a disaster if the team fails to register new or inherent risks on-time, especially in big projects.

But what if you could use an intelligent, self-learning machine to analyze historical data, issue **logs** and incoming requests to come up with an improved risk rating system? Machine learning allows computers to use project data and sophisticated algorithms for predicting outcomes and identifying possibilities of threats and **vulnerabilities** affecting your project. Machine learning can reduce the project failure rate, and you will see significant resources saved.

2. More accurate project planning with AI-enabled process automation

Generally speaking, project planning is about creating a detailed prediction of how to best use the available resources in terms of the project objectives and goals.

As a project manager, you have to re-estimate durations, costs and progress many times throughout the course of your project to make sure it is being performed as planned and to figure out what actions can be taken to make the project happen. The challenge here is that you may perform your estimations in haste and using conservative methods that lead to inflated costs or wrong durations.

As a recurring process, project planning can be optimized with machine learning. An AI-powered system could analyze historical data, productivity rates, time estimates, working hours etc. for patterns to come up with an optimized model of the project management process and automate repetitive tasks. Machine learning could allow for intelligent process automation where computers perform routine tasks. Humans would then do more critical tasks.

3. Project managers will bring in higher value

The job of a project manager is going to be changed drastically. As said earlier, we expect to see repetitive tasks and routines automated, and project managers will dedicate more time to strategic and tactical thinking and judgement.

A project manager will focus on more value-adding tasks while delegating many project management tasks to intelligent machines. For example, job assignment and scheduling will be handled by computers that will use data mining and predictive analytics to design accurate time lines and assign jobs to appropriate team members.

As more and more effective AI assistants reach most industries, project managers will heavily rely their decisions on intelligent machines or Artificial Intelligence Officers (AIOs) that will recommend what's next, alert to vital events, automate time scheduling, and reply to incoming inquiries from superiors and staff.

Key takeaways

- In the near future, software development projects will be managed with the help of AI which will support planning, Quality Assurance (QA), optimization, architecture design and other stages of the development cycle. AI-driven systems will likely replace most lower-level roles and duties altogether.

- In the near future, project managers will share their roles with an augmented assistant, e.g. an artificial intelligence officer that will provide recommendations and insights.

- Data analysts will delegate first-line analysis tasks to deep learning systems and focus on high-level architecture design and support. AI will be absorbed by most industries where project management is actively applied. Low-level roles and routine tasks will almost be effectively automated and handled by self-learning machines, while strategic decision making and judgement will require people more attention.

- The function of human resource within project management will be improved with AI assistance. We can expect to see increased employee retention and engagement as AI will help human resource managers recruit more competent and appropriate employees for specific project team roles and early identify signs of

Unit 3 Planning the Project 项目策划

reducing job satisfaction.

 Words

script ['skrɪpt] n. 脚本
routine [ruː'tiːn] adj. 常规的，日常的
log [lɒg] n. 日志
vulnerability [ˌvʌlnərə'bɪləti] n. 弱点，脆弱性
recur [rɪ'kɜː(r)] v. 循环
drastically ['dræstɪkli, 'drɑːstɪkli] adv. 彻底地

delegate ['delɪgət, 'delɪgeɪt] v. 委托，授（权），把……委托给他人
superior [suː'pɪəriə(r), sjuː'pəriə(r)] n. 上级，高手
augmented [ɔːg'mentɪd] adj. 增强的
insight ['ɪnsaɪt] n. 洞察力
engagement [ɪn'geɪdʒmənt] n. 参与度

 Phrases

just about   几乎，差不多
figure out   想出，理解，断定
in haste   急忙地，草率地，慌张地
inflated cost   滥计成本，虚列成本
dedicate to   把（时间、精力等）用于
key takeaway   关键点，关键信息，重要信息

 Exercises

Ⅰ. Read the following statements carefully, and decide whether they are true (T) or false (F) according to the text.

　　____ 1. AI-driven systems will likely replace most higher-level roles and duties altogether.
　　____ 2. With AI assistance the function of human resource within project management will be improved.
　　____ 3. Data analysts will delegate second-line analysis tasks to deep learning systems and focus on low-level architecture design and support.
　　____ 4. With machine learning project planning can be optimized.
　　____ 5. More projects fail thanks to predictive analytics.

Ⅱ. Choose the best answer to each of the following questions according to the text.

　　1. How many ways can AI improve project management in the near future?

(A) At least one way

(B) At least two ways

(C) At least three ways

(D) None of the above

2. Which of the following is wrong about machine learning regarding a project?

(A) Machine learning allows computers to use project data and sophisticated algorithms for predicting outcomes and identifying possibilities of threats and vulnerabilities affecting your project.

(B) Machine learning will help replace most higher-level roles and duties altogether.

(C) Project planning can be optimized with machine learning.

(D) Machine learning can reduce the project failure rate.

3. Which of the following is wrong according to this text?

(A) With AI assistance the function of human resource within project management will be improved.

(B) Less projects fail thanks to predictive analytics.

(C) Data analysts will delegate second-line analysis tasks to deep learning systems and focus on low-level architecture design and support.

(D) AI-driven systems will likely replace most lower-level roles and duties altogether.

Ⅲ. **Fill in the blanks with the words or phrases chosen from the box. Change the forms where necessary.**

> have   addition   return   facilitate   implement
> opportunity   careful   however   take   equip

How AI Transform Project Management

AI creates the possibility of automated processes and intelligent tools that will reduce manual work. ___1___, based on our experience it will require a certain degree of project management maturity. Furthermore, for AI to bring deep insights into a project or projects, it needs to be ___2___ with a heavy data set from which it can learn what works and what doesn't. ___3___ large historical data sets and current project information in a standardized form is truly one of the key challenges when it comes to successfully ___4___ an AI-based project management system.

In ___5___, if you want to implement an AI system in your existing project management environment, it's imperative to evaluate what benefits it can actually bring to your projects, as well as your business culture and risk appetite. Do you want simple automation—a digital assistant to ___6___ care of menial tasks for you—or do you need it to be more sophisticated and challenge the project in depth? Finally, you also need to ___7___ evaluate what it will cost to realize these potential benefits.

We see great ___8___ for implementing AI-based project systems in large project organizations and in project portfolio management as a way of ___9___ predictive steering of complex transformation projects and portfolios, and thus boosting project success and ___10___ on investment.

Ⅳ. **Translate the following passage into Chinese.**
Project Manager and AI

AI will assist, not replace, project managers. As with every technology, AI alone will not guarantee success. However, deployed purposefully, AI can be a distinctive accelerator and game changer for project managers and thus help increase project success rates. The project managers who succeed will likely be those who manage to see beyond the bounds of 'human' imagination, and answer questions about how this technology can add real value and drive positive change in project management and business transformations. This will ensure the strategic value of project management.

Part 3

Simulated Writing: Software Project Plan

To communicate risk analysis and management, project cost estimates, schedule, and organization to our customers, we usually write a document called a project plan. The plan puts in writing the customer's needs, as well as what we hope to do to meet them. The customer can refer to the plan for information about activities in the development process, making it easy to follow the project's progress during development. We can also use the plan to confirm with the customer any assumptions we are making, especially about cost and schedule.

A good project plan includes the following items:

(1) project scope

(2) project schedule

(3) project team organization

(4) technical description of the proposed system

(5) project standards, procedures, and proposed techniques and tools

(6) quality assurance plan

(7) configuration management plan

(8) documentation plan

(9) data management plan

(10) resource management plan

(11) test plan

(12) training plan

(13) security plan

(14) risk management plan

(15) maintenance plan

The scope defines the system boundary, explaining what will be included in the system and what will not be included. It assures the customer that we understand what is wanted. The schedule can be expressed using a work breakdown structure, the deliverables, and a timeline to show what will be happening at each point during the project life cycle. A Gantt chart can be useful in illustrating the parallel nature of some of the development tasks.

The project plan also lists the people on the development team, how they are organized, and what they will be doing. As we have seen, not everyone is needed all the time during the project, so the plan usually contains a resource allocation chart to show staffing levels at different times.

Writing a technical description forces us to answer questions and address issues as we anticipate how development will proceed. This description lists hardware and software, including compilers, interfaces, and special-purpose equipment or software. Any special restrictions on cabling, execution time, response time, security, or other aspects of functionality or performance are documented in the plan.

Unit 3　Planning the Project　项目策划

Software Project Plan Template

Software Project Plan

for

< Name of Project >

< author >

< date >

Table of Contents

Build the table of contents here. Insert it when you finish your document.

1. Introduction

This section of the Software Project Plan (SPP) provides an overview of the project.

1.1 Project Overview

Include a concise summary of the project objectives, major work activities, major milestones, required resources, and budget. Describe the relationship of this project to other projects, if appropriate. Provide a reference to the official statement of product requirements.

1.2 Project Deliverables

List the primary deliverables for the customer, the delivery dates, delivery locations, and quantities required satisfying the terms of the project agreement.

1.3 Evolution of the SPP

Describe how this plan will be completed, disseminated, and put under change control. Describe how both scheduled and unscheduled updates will be handled.

1.4 Reference Materials

Provide a complete list of all documents and other sources of information referenced in the plan. Include for each the title, report number, date, author, and publishing organization.

1.5 Definitions and Acronyms

Define or provide references to the definition of all terms and acronyms required to properly interpret the SPP.

2. Project Organization

This section specifies the process model for the project and its organizational structure.

2.1 Process Model

Specify the life cycle model to be used for this project or refer to an organizational standard model that will be followed. The process model must include roles, activities, entry criteria and exit criteria for project initiation, product development, product release, and project termination.

2.2 Organizational Structure

Describe the internal management structure of the project, as well as how the project relates to the rest of the organization. It is recommended that charts be used to show the lines of authority.

2.3 Organizational Interfaces

Describe the administrative and managerial interfaces between the project and the primary entities with which it interacts. Table 3-1 may be a useful way to represent this.

Table 3-1 Organizational Interfaces

Organization	Liaison	Contact Information
Customer: < name >	< name >	< phone, E-mail, etc. >
Subcontractor: < name >		
Software Quality Assurance		
Software Configuration Management		
< etc. >		

2.4 Project Responsibilities

Identify and state the nature of each major project function and activity, and identify the individuals who are responsible for those functions and activities. Table 3-2 may be used to depict project responsibilities.

Table 3-2 Project Responsibilities

Role	Description	Person
Project Manager	Leads project team; responsible for project deliverables	< name >
Technical Team Leader(s)	< define as locally used >	< name >
< etc. >	< etc. >	

3. Managerial Process

This section of the SPP specifies the management process for this project.

3.1 Management Objectives and Priorities

Describe the philosophy, goals, and priorities for managing this project. A flexibility matrix might be helpful in communicating what dimensions of the project are fixed, constrained and flexible. Each degree of flexibility column can contain only one "X" (Table 3-3).

Table 3-3 Flexibility Matrix

Project Dimension	Fixed	Constrained	Flexible
Cost		X	
Schedule	X		
Scope (functionality)			X

3.2 Assumptions, Dependencies, and Constraints

State the assumptions on which the project is based, any external events the project is dependent upon, and the constraints under which the project is to be conducted. Include an explicit statement of the relative priorities among meeting functionality, schedule, and budget for this project.

3.3 Risk Management

Describe the process to be used to identify, analyze, and manage the risk factors associated with the project. Describe mechanisms for tracking the various risk factors and implementing contingency plans. Risk factors that should be considered include contractual risks, technological risks, risks due to size and complexity of the product, risks in personnel acquisition and retention, and risks in achieving customer acceptance of the product. The specific risks for this project and the methods for managing them may be documented here or in another document included as an appendix or by reference.

3.4 Monitoring and Controlling Mechanisms

Define the reporting mechanisms, report formats, review and audit mechanisms, and other tools and techniques to be used in monitoring and controlling adherence to the SPP. Project monitoring should occur at the level of work packages. Include monitoring and controlling mechanisms for the project support functions (quality assurance, configuration management, documentation and training).

A table may be used to show the reporting and communication plan for the project. The communication table can show the regular reports and communication expected of the project, such as weekly status reports, regular reviews, or as-needed communication. The exact types of communication vary between groups, but it is useful to identify the planned means at the start of the project (Table 3-4).

Table 3-4 Communication and Reporting Plan

Information Communicated	From	To	Time Period
Status Report	Project Team	Project Manager	Weekly
Status Report	Project Manger	Software Manager, Project Team	Weekly
Project Review	Project Team	Software Manager	Monthly
<etc.>			

3.5 Staffing Approach

Describe the types of skills required for the project, how appropriate personnel will be recruited, and any training required for project team members.

4. Technical Process

This section specifies the technical methods, tools, and techniques to be used on the project. It also includes identification of the work products and reviews to be held and the plans for the support group activities in user documentation, training, software quality assurance, and configuration management.

4.1　Methods, Tools, and Techniques

Identify the computing system(s), development method(s), standards, policies, procedures, team structure(s), programming language(s), and other notations, tools, techniques, and methods to be used to specify, design, build, test, integrate, document, deliver, modify or maintain the project deliverables.

4.2　Software Documentation

Specify the work products to be built for this project and the types of peer reviews to be held for those products. It may be useful to include a table that is adapted from the organization's standard collection of work products and reviews. Identify any relevant style guide, naming conventions and documentation formats. In either this documentation plan or the project schedule provide a summary of the schedule and resource requirements for the documentation effort.

To ensure that the implementation of the software satisfies the requirements, the following documentation is required as a minimum.

4.2.1　Software Requirements Specification (SRS)

The SRS clearly and precisely describes each of the essential requirements (functions, performances, design constraints, and attributes) of the software and the external interfaces. Each requirement is defined such that its achievement is capable of being objectively verified and validated by a prescribed method, for example, inspection, analysis, demonstration, or test.

4.2.2　Software Design Description (SDD)

The SDD describes the major components of the software design including databases and internal interfaces.

4.2.3　Software Test Plan

The Software Test Plan describes the methods to be used for testing at all levels of development and integration: requirements as expressed in the SRS, designs as expressed in the SDD, code as expressed in the implemented product. The test plan also describes the test procedures, test cases, and test results that are created during testing activities.

4.3　User Documentation

Describe how the user documentation will be planned and developed. (This may be just a reference to a plan being built by someone else.) Include work planned for online as well as paper documentation, online help, network accessible files and support facilities.

4.4　Project Support Functions

Provide either directly or by reference, plans for the supporting functions for the software project. These functions may include, but are not limited to, configuration management, software quality assurance, and verification and validation. Plans for project support functions are developed to a level of detail consistent with the other sections of the SPP. In particular, the responsibilities, resource requirements, schedules and budgets for each supporting function must be specified. The nature and type of support functions required will vary from project to project, however, the absence of a software quality assurance, configuration management, or, verification and validation plan must be explicitly justified in project plans that do not include them.

5. Work Packages, Schedule, and Budget

Specify the work packages, dependency relationships, resource requirements, allocation of budget and resources to work packages, and a project schedule. Much of the content may be in appendices that are living documents, updated as the work proceeds.

5.1 Work Packages

Specify the work packages for the activities and tasks that must be completed in order to satisfy the project agreement. Each work package is uniquely identified. A diagram depicting the breakdown of project activities and tasks (a work breakdown structure) may be used to depict hierarchical relationships among work packages.

5.2 Dependencies

Specify the ordering relations among work packages to account for interdependencies among them and dependencies on external events. Techniques such as dependency lists, activity networks, and the critical path method may be used to depict dependencies among work packages.

5.3 Resource Requirements

Provide, as a function of time, estimates of the total resources required to complete the project. Numbers and types of personnel, computer time, support software, computer hardware, office and laboratory facilities, travel, and maintenance requirements for the project resources are typical resources that should be specified.

5.4 Budget and Resource Allocation

Specify the allocation of budget and resources to the various project functions, activities, and tasks.

5.5 Schedule

Provide the schedule for the various project functions, activities, and tasks, taking into account the precedence relations and the required milestone dates. Schedules may be expressed in absolute calendar time or in increments relative to a key project milestone.

6. Additional Components

Certain additional components may be required and may be appended as additional sections or subsections to the SPP. Additional items of importance on any particular project may include subcontractor management plans, security plans, independent verification and validation plans, training plans, hardware procurement plans, facilities plans, installation plans, data conversion plans, system transition plans, or the product maintenance plan.

6.1 Index

An index to the key terms and acronyms used throughout the SPP is optional, but recommended to improve usability of the SPP.

6.2 Appendices

Appendices may be included, either directly or by reference, to provide supporting details that could detract from the SPP if included in the body of the SPP. Suggested appendices include:

A. Current Top 10 Risk Chart
B. Current Project Work Breakdown Structure
C. Current Detailed Project Schedule

P.S. This template can be tailored according to the different projects.

A Sample Software Project Planning

Software Project Plan
For
Web Publishing System

Version 1.0

Victor Lu (Team leader)
Jack Liang
Jason Wang
Mary Zeng
David Chang

April 7, 2020

Table of Content

1. Introduction
   1.1 Scope of project
   1.2 Major software functions
   1.3 Major constraints
   1.4 Purpose of document
   1.5 Project deliverables
   1.6 Glossary
   1.7 References
2. Project Organization
   2.1 Project schedule
   2.2 Deliverables and milestones
   2.3 Project estimation chart
   2.4 People (project group)/ Staff organization
   2.5 Management reporting and communication
3. Tracking and Control Mechanisms
   3.1 Quality assurance and control
   3.2 Change management and control
4. Method and Tool
5. Risk Management

1. Introduction

1.1 Scope of project

This software system will be a Web Publishing System for a local editor of a regional historical society. It will be designed to maximize the editor's productivity by providing tools to assist in automating the article review and publishing process, otherwise it has to be performed manually. By maximizing the editor's work efficiency and production, the system will meet the editor's needs while remaining easy to understand and use.

1.2 Major software functions

Eventually, editor could use this product to manage and communicate with a group of reviewers and authors to publish articles more efficiently. Moreover, the software could facilitate communication between authors, reviewers, and the editor via E-mails.

1.3 Major constraints

- Budget: ¥150 000
- Time: two months
- Staff: 5 members in this group.

1.4 Purpose of document

This document describes the plan of the Web Publishing System project. It also provides cost, effort and time estimate of this project which is used to monitor and control the project to guide the project execution. The aim of these processes is to ensure the developers complete the project within budget as well as by the due date.

1.5 Project deliverables

All items listed in this subsection are the deliverables requested by the Web Publishing System.

Software documentation:
- Installation documentation
- End-user documentation

Project documentation:
- Software Project Plan (SPP)
- Software Requirements Specification (SRS)
- Software Design Specification (SDS)
- Software Test Plan
- Software Testing Report

1.6 Glossary

Glossary is as shown in Table 3-5.

Table 3-5 Glossary

Term	Definition
Author	A person submitting an article to be reviewed. In case of multiple authors, this term refers to the principal author, with whom all communication is made
Database	A collection of information monitored by this system
Editor	A person who receives articles, sends articles for review, and makes final judgments for publications
Historical Society Database	The existing membership database (also HS database)
Member	A member of the Historical Society listed in the HS database
Reader	Anyone visiting the site to read articles
Review	A written recommendation about the appropriateness of an article for publication; may include suggestions for improvement
Reviewer	A person that examines an article and has the ability to recommend approval of the article for publication or to request that changes be made in the article

1.7 References

IEEE Standard for Software Project Management Plans (IEEE 1058—1998).

2. Project Organization

2.1 Project schedule

This section presents an overview of project tasks and schedule. The project will use the waterfall development approach for fulfilling the tasks of different phases on time.

2.2 Deliverables and milestones

Deliverables and milestones are as shown in Table 3-6.

Table 3-6 Deliverables and Milestones

Stage Of Development	Stage Completion Date	Milestone	Deliverable Completion Date
Planning	04/07/20	Project Plan completed	04/07/2020
Requirements Definition	04/10/20	Draft Requirements Specification completed	04/08/2020
		Draft Design Specification completed	04/09/2020
		Requirements Specification (final) completed	04/10/2020
Design (Functional & System)	04/22/20	Draft Testing Plan completed	04/15/2020
		Program and Database Specifications completed	04/17/2020
		Design Specification (final) completed	04/22/2020
Programming	05/12/20	System Test Plan completed	05/8/2020
		Software (frontend and backend) completed	05/12/2020
Integration & Testing	05/21/20	System Test Plan (final) completed	05/14/2020
		Test Reports completed	05/18/2020
		User's Guide completed	05/21/2020

continued

Stage Of Development	Stage Completion Date	Milestone	Deliverable Completion Date
Installation & Acceptance	06/02/20	Maintenance Plan completed	05/25/2020
		Project Closeout	06/02/2020

2.3 Project estimation chart

Table 3-7 illustrates the project estimation schedule. Figure 3-3 illustrates the detailed project estimation schedule and Figure 3-4 is Gantt chart.

Table 3-7 Project Estimation Schedule

Task Name	Duration	Staff
Allocate project resources	2 days	1
Establish project environment	2 days	2
Create software project plan	1 day	1
Define and develop software requirements	2 days	1
Create Software Requirements Specification(SRS)	1 day	1
Perform architectural design	3 days	2
Design the interfaces	2 days	1
Design the database	2 days	1
Create Software Design Specification(SDS)	1 day	3
Programming	14 days	3
Plan testing	2 days	2
Execute the tests	3 days	2
User manual development	2 days	2
Reapply another software lifecycle	8 days	4

		任务模式	任务名称	工期	开始时间	完成时间	前置任务	资源名称
1			▲Project Initiation	4 个工作日	2020年4月1日	2020年4月6日		
2			Allocate project resources	2 个工作日	2020年4月1日	2020年4月2日		Victor
3			Establish project environment	2 个工作日	2020年4月3日	2020年4月6日		David, Victor
4			▲Planning	1 个工作日	2020年4月7日	2020年4月7日		
5			Create Software Project Plan	1 个工作日	2020年4月7日	2020年4月7日		Victor
6			▲Requirement	3 个工作日	2020年4月8日	2020年4月10日		
7			Define and develop software requirement	2 个工作日	2020年4月8日	2020年4月9日		Jack
8			Create Software Requirements Specification	1 个工作日	2020年4月10日	2020年4月10日	7	Jack
9			▲Design	8 个工作日	2020年4月13日	2020年4月22日		
10			Perform architectural design	3 个工作日	2020年4月13日	2020年4月15日	8	Victor, Mary
11			Design interfaces	2 个工作日	2020年4月16日	2020年4月17日	8	Jason
12			Design the database	2 个工作日	2020年4月20日	2020年4月21日	8	Victor
13			Create Softmare Design Specification	1 个工作日	2020年4月22日	2020年4月22日		Mary, Jason, David
14			▲Implementation	14 个工作日	2020年4月23日	2020年5月12日	10, 11, 12	
15			Create source code	14 个工作日	2020年4月23日	2020年5月12日		Mary, Jason, Victor
16			▲Testing	5 个工作日	2020年5月13日	2020年5月19日	5, 8, 13	
17			Create Software Test Plan	2 个工作日	2020年5月13日	2020年5月14日		Mary, Victor
18			Excute manual development	3 个工作日	2020年5月15日	2020年5月19日	17	Mary, David
19			▲Installation	2 个工作日	2020年5月20日	2020年5月21日		
20			User manual	2 个工作日	2020年5月20日	2020年5月21日		Victor, Jack
21			▲Maintenance	8 个工作日	2020年5月22日	2020年6月2日		
22			Reapply another software lifecycle	8 个工作日	2020年5月22日	2020年6月2日		Mary, Victor, David, Jason, Jack

Figure 3-3 Detailed Project Estimation Schedule

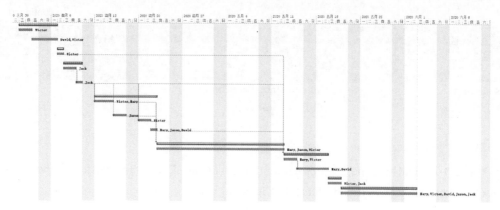

Figure 3-4　Gantt Chart

2.4　People (project group) / Staff organization

Project Manager:

• Victor Lu

E-mail: AAAA@gmail.com

Team members:

• Jack Liang

E-mail: BBBB@gmail.com

• Jason Wang

E-mail: CCCC@gmail.com

• Mary Zeng

E-mail: DDDD@gmail.com

• David Chang

E-mail: EEEE@gmail.com

2.5　Management reporting and communication

Mechanisms for progress reporting and inter/intra team communication are identified.

Meetings

In order to monitor the progress of the project, the project group will be subjected to weekly meetings along with some informal meetings depending on needs. During the weekly meetings, team members will discuss issues and progress through face-to-face meetings and all conclusions and decisions will be documented in the meeting reports. The informal meetings will be called throughout the entire project on an as-needed basis which will be determined by either team members or the clients.

E-mails

E-mails will be the daily communication channel for solving instant problems and confusions. E-mails will also be used to schedule weekly meetings and to transfer documents.

Status Report

Each week, every member has to fill in a working-hour table for that week. The working-hour table separates the whole project into six stages. With the help of the working-hour table, both project manager and team members will be able to keep track of the time spent on the project.

3. Tracking and Control Mechanisms

Techniques to be used for project tracking and control are identified. In order to ensure the project goes well and the specifications are followed, the project team will set some monitoring policy.

3.1 Quality assurance and control

- Tight change management.
- Quality review in a meeting format.
- Extensive before implementation design using rapid prototyping.
- Close contact with clients, meeting every two weeks and regular E-mail contacts.

3.2 Change management and control

- For changes that affect the user experiences we will have to notify all clients.
- For changes that do not affect the user experiences we will notify a client representative.
- Due to the size of the team, internal control panel will be used. If one member of the team suggests a change, it will need to be approved by the other two members.
- Formal version numbering will be used. All version changes must be documented in a common document accessible to all team members before a new version number can be released. Version number will be structured as follows:

＜Major Release＞.＜Minor Release＞＜Bug fix＞

Version changes will be reviewed. Not only the previous version, but also all older versions of any document or codes will be preserved. This will ensure if the need to revert back to more than one version arises, the necessary version is available.

Change management policy

Once a work product has been finalized and approved, all changes to that work product must be submitted through the Git, where the changes will be reviewed and either approved or denied by the project manager, based on the risk profile and perceived benefit of the change to be made. Since the waterfall methodology is being used for this project, requests for changes will be treated conservatively as they will potentially be extremely disruptive to the activities downstream of the change.

4. Method and Tool

Development methodology

The project shall use the waterfall software development methodology to deliver the software products, with work activities organized according to a tailored version of those provided by the IEEE Standard for Developing Software Life Cycle Processes (IEEE 1074—1997).

The Software Project Plan (SPP) will be based on the IEEE Standard for Software Project Management Plans (IEEE 1058-1998).

Tools

The following work categories will have their work products satisfied by the identified tools:

Document Publishing:
- Microsoft Word 2017

Project Management:
- Microsoft Project 2016

Implementation:
- Microsoft Visual Studio 2017
- Microsoft SQL Server 2016
- Web Standards Update for Visual Studio & JScript Editor Extensions

Version Control:
- Git

5. Risk Management

The following part describes all the risks this project team might come up with.

- Team member leaves

Team member may leave because of personal reasons.

How to avoid: Make sure every member understands the importance of the project and has the same determination to achieve it.

Minimize the risk: Make a backup plan. If a team member really leaves, the rest of the team can still complete the project. Before a team member's leaving, he/she should document all work he/she has done, so that other members can take it over.

- Develop a program that is not usable

The resulted program may not meet the client's requirements or cannot run on the assigned platform.

How to avoid: Perform system analysis and system design carefully before writing the program. Do the checkpoints carefully.

Minimize the risk: Discuss with the lecturer, team members and client representatives and see if this can be resolved by changing the design or implementation.

- Incompetent project management

Incompetent project management may lead to the failure of the project.

How to avoid: The project manager will remain up to date with the latest technology and keep track of the project development.

Minimize the risk: Discuss with the lecturer and team members and try to reorganize the team.

- Lack of knowledge

Team members may lack technical or domain knowledge of the project.

How to avoid: All team members have to study hard beforehand, and the client should provide learning resources or contact channels for the development team.

Minimize the risk: Ask for help from experts.

- Time related issues

There may not be enough time for completing this project.

How to avoid: Conduct good time control and monitor all the time.

Minimize the risk: Discuss with the lecturer and the client representatives to see if it is possible to either reduce the scale of the project or extend the developing time.

- Too ambitious requirements

The client may ask for too many functions for this project.

How to avoid: Perform system analysis and system design carefully before writing the program. If there are too ambitious requirements, discuss them with the lecturer and the client representatives.

Minimize the risk: Discuss them with the lecturer and client representatives. Delay the implementation of these requirements till next version or reduce the requirements.

- Bad design

It can include: bad system design or user interface design or working flow design.

How to avoid: Analyze the system and the design properly at the beginning; consult the expert, lecturer and client representatives for their opinions. Review and integrate the system all the time.

Minimize the risk: Consult the experts and lecturer. If there is still enough time, redesign and re-implement the code.

- Lack of client interaction

The project team is possible to lose the real time feedback from the client due to the lack of communication and interaction. This may lead to unusable or unqualified program.

How to avoid: Conduct review meetings and status reports regularly and make sure there is a contact channel between the project team and the client.

Minimize the risk: Visit the client regularly.

- Lack of hardware resource

The project team may not have enough hardware for developing this project, such as test machines.

How to avoid: Prepare the hardware according to the system analysis.

Minimize the risk: Discuss this problem with the client and lecturer.

- Code theft

The code may be stolen during the developing time.

How to avoid: Apply permission control to all the code written, use version control system to control the code, and all code should have at least one backup.

Minimize the risk: Update the backup regularly and firewall the system.

 Exercises

1. What is this plan mainly talking about?
2. What is the main constraint of this project?
3. What are the responsibilities of project manager in this project?

Unit 4

Working in a Team

团队合作

Part 1

Listening & Speaking

Unit 4

Dialogue: Team Structure

Jason: Considering our situation, we can try eXtreme Programming (XP) in the development process[1], which **advocates** communication without any obstacles and leads to a quick **delivery**. It is very appropriate to the characteristics of our team and project.

[1] Replace with: We can try an agile programming in the development process.

Sharon: You are right. With XP, we can cooperate with each other and focus our time and energy, which is needed to succeed.

Kevin: As I know, there is another team structure which **recognizes** that there are three main task categories in software development — management related, development related, and testing related. In this way, we can clearly divide our work into those three main aspects. Jason mainly focuses on development, Sharon conducts testing and I'm responsible for the management of our project, so that we can carry out the task more professionally.

Sharon: Sounds good! Well, let's do it!

Jason: Besides, in our daily work, good communication between members of a development group is essential. We must exchange information on the status of **respective** work, the design decisions that have been made and changes to previous decisions that are necessary.

Sharon: To accomplish this, we had better establish some mechanisms for formal and informal communication among us. Formal communication can be accomplished

Sharon: through writing documents and progress reports. On the other hand, we can share ideas via Internet at any time informally.

Kevin: In addition, since we develop **synchronously** and cooperatively, we need a configuration management mechanism in order to ensure the consistency of our work.

Jason: Yes. Maybe some automatic tools can help us with that.

Sharon: Git is one of the industry's most widely used version control systems [2], and is very **fit for** the size and characteristics of a development team like ours. We can give it a try.

[2] Replace with: SVN is an open-source revision control system

Kevin: Ok, I agree.

Exercises

Work in a group, and make up a similar conversation by replacing the statements with other expressions on the right side.

Words

advocate ['ædvəkeit,-ət] v. 提倡,主张
delivery [di'livəri] n. 交付
recognize ['rekəgnaiz] v. 认可,承认
respective [ri'spektiv] adj. 分别的,各自的

synchronously ['siŋkrənəsli] adv. 同时地,同步地

Phrases

fit for 适合

Listening Comprehension: Project Team

Listen to the article and answer the following 3 questions based on it. After you hear a question, there will be a break of 15 seconds. During the break, you will decide

105

which one is the best answer among the four choices marked (A), (B), (C) and (D).

Questions

1. Which point is not mentioned about the advantages of the team spirit for a group?

    (A) Observing a group quality standard self-consciously.

    (B) Group members working closely together and learning from each other.

    (C) Making every group member unique with a sense of "elite".

    (D) Practicing egoless programming.

2. How many vivid terms describing the team spirit of a group are referred in this article?

    (A) One

    (B) Two

    (C) Three

    (D) Not referred

3. Which is the most appropriate way of communication when making a most important change in design decisions?

    (A) The project manager makes the decision and informs other group members with an oral agreement.

    (B) Members working together share, negotiate and make the decision.

    (C) The system designer makes the decision and explains it to everyone involved in his daily report.

    (D) The whole group holds a meeting to discuss the decision, and write a formal report on it for the configuration management.

 Words

motivate[ˈməutiveit] v. 激发……的积极性	jell[dʒel] n. & v. 结合,相处融洽
cohesiveness[kəuˈhiːsivnəs] n. 内聚性,黏结性	knit[nit] v. 结合
well-led 领导得好的	juggernaut[ˈdʒʌɡənɔːt] n. 巨无霸,世界主宰,强大的破坏力
consensus[kənˈsensəs] n. (意见等)一致	momentum[məˈmentəm] n. 势头,动力,冲力
egoless[ˈiːɡəulis] adj. 不自负的,非自我的	

Unit 4　Working in a Team　团队合作

Phrases

identify with　视……为一体,认同
cope with　对付,应付
go way up　大幅上升

Dictation: Agile Software Development

This article will be played three times. Listen carefully, and fill in the numbered spaces with the appropriate words you have heard.

The modern business environment that **spawns** computer-based systems and software products is ___1___ and **ever-changing**. ___2___ conditions change rapidly, end-user needs ___3___, and new competitive threats ___4___ without warning. In many situations, we no longer are able to define requirements fully before the project begins. Software engineers must be agile ___5___ to respond to a **fluid** ___6___ environment.

The modern definition of agile software development evolved in the mid-1990s. ___7___, agile methods were called "___8___ methods" as part of a reaction against "heavyweight" methods. The aim of it is to create software in a lighter, faster, more ___9___ way. In 2001, several of the most **prominent proponents** of those "lightweight methodologies" started the Agile ___10___ and signed the "**Manifesto** for Agile Software Development", a ___11___ of the values shared by them, for those **contemplating** new agile development processes, that is: individual and ___12___ over processes and tools, working software over comprehensive ___13___, ___14___ collaboration over contract negotiation, ___15___ to change over following a plan.

Actually, a number of methods similar to Agile were created ___16___ to 2000. ___17___ Programming (XP) is the most widely used agile process created by Kent Beck in 1996. An **adaptive** software development process was introduced in a paper by Edmonds in 1974. Other **notable** agile methods include Scrum, Crystal, ___18___ Software Development, ___19___ Driven Development, and ___20___ Systems Development Method.

Words

spawn[spɔːn] v. 大量产生,造成,引发,引起　　ever-changing 千变万化的,常变的

107

fluid [ˈfluːid] *adj.* 变化的，流动的，液体的
prominent [ˈprɔminənt] *adj.* 著名的，重要的
proponent [prəˈpəunənt] *n.* 支持者，倡导者
manifesto [ˌmæniˈfestəu] *n.* 宣言，声明
contemplate [ˈkɔntempleit] *v.* 思考，预期
adaptive [əˈdæptiv] *adj.* 适应的
notable [ˈnəutəbl] *adj.* 著名的

Part 2

Reading & Translating

Section A: Team Structure

We have seen that the number of resources is fixed when schedule is being planned. Detailed scheduling is done only after actual assignment of people has been done, as task assignment needs information about the capabilities of the team members. In general, we implicitly assume that the project's team is led by a project manager, who does the planning and task assignment. This form of hierarchical team organization is fairly common, and was earlier called the Chief Programmer Team.

In this hierarchical organization, the project manager is responsible for all major technical decisions of the project. He does most of the design and assigns coding of the different parts of the design to the programmers. The team typically consists of programmers, testers, a configuration controller, and possibly a librarian for documentation. There may be other roles like database manager, network manager, backup project manager, or a backup configuration controller. It should be noted that these are all logical roles and one person may do multiple such roles.

For a small project, a one-level hierarchy suffices. For larger projects, this organization can be extended easily by partitioning the project into modules, and having module leaders who are responsible for all tasks related to their module and having a team with them for performing these tasks.[1]

A different team organization is the egoless team. Egoless teams consist of ten or fewer programmers. The goals of the group are set by consensus, and input from every member is taken for major decisions. Group leadership rotates among the group members. Due to their nature, egoless teams are sometimes called democratic teams. This structure allows input from all members, which can lead to better decisions for difficult problems. This structure is well suited for long-term research-type projects that do not have time constraints. It is not suitable for regular tasks that have time constraints; for such tasks, the communication in democratic structure is unnecessary and results in inefficiency.

In recent times, for very large product developments, another structure has emerged (Figure 4-1). This structure recognizes that there are three main task categories in software development—management related, development related, and testing related. It also recognizes that it is often **desirable** to have the test and development team be relatively independent, and also not to have the developers or testers report to a non-technical manager.[2] In this structure, consequently, there is an overall unit manager, under whom there are three small hierarchic organizations — for program management, for development, and for testing. The primary job of developers is to write code and they work under a development manager. The responsibility of the testers is to test the code and they work under a test manager. The program manager provides the specifications for what is being built, and ensures that development and testing are properly coordinated. In a large product this structure may be replicated, one for each major unit. This type of team organization is used in corporations like Microsoft.

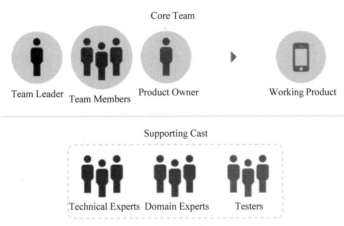

Figure 4-1　A Team Structure

 Words

fix[fiks] v. 确定,处理,解决	backup['bækʌp] adj. 候补的,备份的
capability[ˌkeipə'biləti] n. 才能,能力	suffice[sə'fais] v. 足够,满足
implicitly[im'plisitli] adv. 暗含地,含蓄地	partition[pɑː'tiʃn] v. 划分,把……分成部分
hierarchical[ˌhaiə'rɑːkikl] adj. 分等级的,层次的	rotate[rəu'teit] v. 轮流,交替
fairly['feəli] adv. 相当地;公平合理地	desirable[di'zaiərəbl] adj. 可取的,值得做的

Phrases

be suited for　适合于

Notes

[1] **Original**: For larger projects, this organization can be extended easily by partitioning the project into modules, and having module leaders who are responsible for all tasks related to their module and having a team with them for performing these tasks.

Translation: 对于较大的项目,通过将项目划分成模块,并且拥有模块领导,他们负责与其模块相关的所有任务并带领一支执行这些任务的团队,这样的组织能够很容易地被扩展。

[2] **Original**: It also recognizes that it is often desirable to have the test and development team be relatively independent, and also not to have the developers or testers report to a non-technical manager.

Translation: 这种结构还认为,让测试和开发团队相对独立,以及不让开发人员或测试人员向非技术性经理汇报,通常可以收到令人满意的效果。

 Exercises

Ⅰ. Read the following statements carefully, and decide whether they are true (T) or false (F) according to the text.

____ 1. Both the hierarchical team structure and the team structure used in Microsoft are extensible.

____ 2. Group leadership is not fixed in the egoless team.

____ 3. One person must act a specific and single role in the hierarchical team.

____ 4. Among all the team structures mentioned in this article, it is only in the Chief Programmer Team that a hierarchical structure is used.

____ 5. In general, there are three main task categories in software development.

Ⅱ. Choose the best answer to each of the following questions according to the text.

1. Which of the following statements is wrong according to this article?

(A) This article refers to three types of team structures in all.

(B) Hierarchical team organization is used the most frequently.

(C) Because of the project manager being responsible for all major technical decisions, the communication in the hierarchical organization is unnecessary and results in inefficiency.

(D) There is no the best team structure but only the most suitable one for a specific project.

2. Which of the following figures does properly present the third team structure

mentioned in this article?

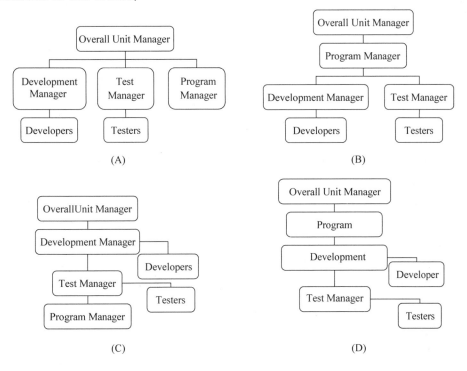

3. Suppose that you are going to develop a long-range exploratory project with some new techniques, which of the following team structures is the most suitable for you?

(A) One-level hierarchical team.

(B) Extended hierarchical team.

(C) Democratic team.

(D) The team organization which is used in Microsoft, mentioned in the last paragraph.

Ⅲ. Fill in the numbered spaces with the words or phrases chosen from the box. Change the forms where necessary.

> like   reach   enough   hand   resolve
> effective   create   critical   define   work

Evaluation of Team Performance

Let's say your team is working, and now it's time to see how good, how ___1___ it actually is. Any and all evaluation of team performance should focus on two things: team results and team process.

Team results are ___2___ by the team's main objectives. It may be anything—from better product quality, to faster delivery time, to less resources used. Team process, on the other ___3___, is the way the team goes about achieving results—how well team members ___4___, share information, manage budgets, schedules, and interpersonal relations.

If you see how well your team ___5___, an evaluation will allow you to set markers for future reference. But if a team faces internal obstacles, a detailed evaluation of all the team processes becomes even more ___6___. There are numerous methods you can use to evaluate your team's processes, but the most simple and effective include benchmarking, ongoing team discussions, and project debriefings.

As you see, building a software development team is a bit ___7___ putting together a baseball team. Everyone has and knows their role, their position on the field, and the goal everyone's longing for. It's not ___8___ just to pick out good professionals; they have to match each other like puzzle pieces. And it's your job to really get to know each candidate to see how they work together. Learn how to appreciate teamwork, how to celebrate small achievements on your way to ___9___ the big goal. Create a dedicated development team that's going to be successful in this particular constellation. Better yet, build a team that will want to work together even after your project is done—this will be a true sign that you ___10___ not just a good outsourced development team, but a great one.

Ⅳ. Translate the following passage into Chinese.

How to Form a Successful Development Team（1）

A good team isn't something that happens on its own. But why do we need one at all? Don't good professionals form a good team by default? The thing is, it's not enough to lock them in a room and give them a deadline. An efficient team isn't just about professionalism; it's also about how the team members interact. Only a truly efficient team can deal with the workload faster and be more productive. And generally, it's always nice to work with a team that, well, doesn't mess up.

Section B：How to Build a Highly Effective Software Development Team

Software development is both a science and a process. If you are a project manager in charge of creating a software solution for a client, then you must have a team of professionals who would work with you towards a common goal. You have a creative team highly experienced in what they do and they *are* all *geared to* produce a solution that no one else has developed before. How do you *go about* it?

Communication is the key for building a highly effective software development team. A highly effective team would provide value to a project, to a company and they remain productive even in *adverse* situations (Figure 4-2). When a team feels it has the support of its project manager, there is motivation and results will be naturally generated. That is why in every team, communication is the key; when you know how to communicate to your team internally and externally, success follows.

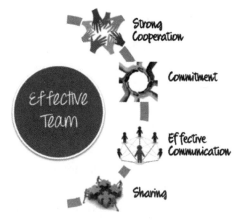

Figure 4-2　An Effective Software Development Team

While planning a communication strategy with your team, it is **imperative** how often you hold meetings. The meetings that you conduct must be qualitative. There is **no point** in holding meetings twice or **thrice** a day if it affects the working hours and productivity of your team members. While conducting meetings, it is important to **cultivate introspection**. Do a communication review—are you conducting a group meeting when all you need to do is have a discussion with just two or three members of the team? Isn't it better to call just the necessary members of the team than call everyone? That way you can leave the rest to work on the project. Once the meeting with the selected few are over, you can send E-mails to the rest of the team and inform them of any developmental changes to the project. When you need to do a performance review, there is no need to call everyone in the team until absolutely necessary. Hence, when you communicate with your team, there are three basic things to remember

- Communicate less
- Communicate qualitatively
- Communicate efficiently

Planned meetings are scheduled interruptions and if the meeting likely bring **actionable results**, then it is a waste of time. In order to ensure you don't waste a single minute in the meeting, have chart of what is to be discussed, the list of questions to be asked and main objectives of the meeting.

Once the problem of communication is solved, you can plan for shorter work plans. Every project has milestones and when you **cut shot** the milestones and ensure everyone in the team deliver complete **deliverables as per** the milestone, there is success. You can **shorten** workplaces and make it easy for the rest of your team **chart** their own strategies, and even take risks. When you divide the project into different milestones, you can always go back and retrace the steps if you **come across** a hurdle. That way you don't lose a lot of time trying to figure out the source of the problem.

If it is a team, it should behave like a team, which means there must be interdependence. There must be shared responsibilities and people with similar skills must be aware of each other's tasks so they both won't do the same thing and lose time. A team would share concerns, ideas and responsibilities so the project would continue as scheduled, and everybody would **do their bit**.

In order to successfully manage your team, you must be able to **instill** trust and respect among team members. If you are getting together a team of software professionals who have the reputation of not **getting along with** each other, then you must either be able to **work** it **out** with them and find out what is causing this friction or you must **dissolve** the team because there is going to be no progress without trust or respect. Trust takes time to build, but if it is among team members who have had a past, then it is going to take a longer time. If it is not possible to dissolve the team, then you will have to call everyone together, and see if it is possible to remove the friction.

Agile Methodology is the most successful feature driven development for project managers. Being agile helps team members to respond to unpredictability in a project. Scrum [1] is another term used by project managers when they use agility in their project management. Thanks to the popularity of Scrum in Agile Methodology, many project managers have adopted this method of doing projects. Agile software development is a process through which project managers and team members get their work done. It contains a set of guidelines through which each team would be able to deliver maximum deliverable at minimum production cost. Agile methods have enabled effective workflow management and gets projects done faster. Agile Methodology has made it possible for team members to work successfully on complex software development projects.

It could be challenging for team members who are not used to agile working techniques to get used to it the first time. Project managers must learn to break their old habits if they want complete success with Agile. They must know what to do with processes and procedures that **stand in the way** of productivity. Only if the team works as one whole unit will they be able to **commit** successfully **to** the processes.

Words

adverse['ædvɜ:s, əd'vɜ:s] *adj.* 不利的, 相反的, 敌对的
imperative[im'perətiv] *adj.* 重要的, 必要的
thrice[θrais] *n.* 三次, 三倍
cultivate['kʌltiveit] *v.* 培养, 陶冶
introspection[,intrə'spekʃn] *n.* 内省, 反省

deliverable [di'livərəbl] *n.* 应交付的产品
shorten['ʃɔ:tn] *v.* 减少
chart[tʃɑ:t] *v.* 详细计划, 记录
instill[in'stil] *v.* 逐步灌输
dissolve[di'zɔlv] *v.* 解散

Unit 4　Working in a Team　团队合作

 Phrases

be geared to do　准备好做
go about　着手做
no point　毫无意义, 没理由, 无济于事
actionable results　有目共睹的结果
cut shot　缩短, 打断, 缩减
as per　按照, 依据, 如同
come across　偶遇, 无意中发现
do one's bit　尽自己的一份力量
get along with　与……和睦相处, 取得进展
work out　解决, 实现
stand in the way　碍事, 挡道
commit to　使(自己)致力于(某事或做某事)

 Notes

［1］ Scrum 是迭代式增量软件开发过程, 通常用于敏捷软件开发。Scrum 是一个实现了敏捷思维的框架, 能够帮助团队快速前进和学习, 是一种事半功倍进行团队协作的敏捷工作方法, Scrum 常常与其他敏捷框架和实践(例如极限编程)结合起来使用。

 Exercises

Ⅰ. Read the following statements carefully, and decide whether they are true (T) or false (F) according to the text.

　　＿＿＿1. You must be able to instill trust and respect among team members if you manage your team successfully.
　　＿＿＿2. Thanks to the popularity of Waterfall Model in Agile Methodology, many project managers have adopted this method of doing projects.
　　＿＿＿3. Project managers must know what to do with processes and procedures that stand in the way of productivity.
　　＿＿＿4. Communicate less, communicate qualitatively and communicate efficiently are three basic things to remember when you communicate with your team.
　　＿＿＿5. It is meaningful in holding meetings twice or thrice a day if it affects the working hours and productivity of your team members.

Ⅱ. Choose the best answer to each of the following questions according to the text.

1. Which of the following should be remembered when you communicate with your team?

   (A) Communicate less

   (B) Communicate qualitatively

   (C) Communicate efficiently

   (D) All of the above

2. Which of the following belongs to Agile Methodology?

   (A) Waterfall Model

   (B) Incremental Process Model

   (C) Scrum

   (D) None of the above

3. Which of the following statements is wrong about Agile Methodology?

   (A) It could be challenging for team members who are not used to agile working techniques to get used to it the first time.

   (B) Project managers must learn to break team members' old habits if they want complete success with Agile.

   (C) Agile Methodology has made it possible for team members to work successfully on complex software development projects.

   (D) None of the above.

Ⅲ. Fill in the numbered spaces with the words or phrases chosen from the box. Change the forms where necessary.

> include   accept   to   communicate   into
> prefer   while   aim   from   on

Dedicated Development Teams: Psychological Syncing

Don't forget that all of the team members are just humans. This means, apart ___1___ professional skills, you should take ___2___ account psychological factors. Based ___3___ individual psychological characteristics, a developer team is effective when:

- The size is minimal (3 to 10 people) while ___4___ a full skill set for a particular project time frame;
- The team core is very cohesive, but lack of ___5___ with certain team members is possible (outsourcing of minor tasks);

- The leader is highly ___6___ (by at least half of the team), while the member with the highest acceptance is the one with the highest level of intelligence that allows him or her to make the biggest contribution;
- The leaders' team is aimed at itself, the team members are ___7___ at the goal, the number of members aimed at communication is minimal;
- The team leader aims to serve, ___8___ the leaders' team's aim is mastery;
- The team has a full set of roles: idea generator, analyst, and critic. Acting roles preside over mental roles, while mental roles preside over social ones. The number of idea generators and analysts isn't large compared ___9___ other roles;
- The intellectual capital of the team should have different levels: a combination of one intellectual leader with several social leaders with lower intellectual capital is ___10___.

Ⅳ. **Translate the following passage into Chinese.**

How to Form a Successful Development Team (2)

Statistics show that, in fact, the main reason why projects fail is a lack of confidence in the project's success: "75% of respondents admit that their projects are either always or usually doomed right from the start." But why do you think it happens, and how can you change it? In order to function properly, the team needs to know all the aspects of the process, their duties and responsibilities, and believe in what they do—and you're the one to convince them. Here's how to build a software development team you can rely on, an effective team you're confident about, one that you can look at and say, "We're going to change the world!"

Part 3

Simulated Writing: PowerPoint Presentation

Introduction

A PowerPoint presentation is similar to a poster presentation, only the information is on computer slides rather than actual posters. They are usually used to accompany an oral presentation; they should enhance the oral presentation. You can incorporate audio and visual media. They are often used to share information with a large group, such as at a professional conference, classroom presentations, and meetings. It should be more like an outline of your presentation. There are three main elements to a PowerPoint presentation: text, images, and tables or graphs. Text allows you to reinforce main points as well as key terms and concepts. Images illustrate or highlight main points. Tables and graphs present information in a way that is easy to understand and see.

Some Preliminaries

Before start, you should consider several key parts of your project: your audience, purpose (persuasive, informative, etc.), subject matter and presentation. Since good PowerPoint projects come from making design decisions that fit the occasion, knowing as much as possible about the rhetorical situation before you create your PowerPoint will ensure your success.

Common Components

The slides for a PowerPoint presentation should be more like an outline. Text is often listed rather than written in full sentences. The following are a few of the things that can be presented on a PowerPoint slide:
- Graphs and/or tables
- Definitions
- Lists
- Essential facts
- Necessary images

Arrangement

The order of your PowerPoint will depend on what you think your audience needs. Whatever you do, organize your PowerPoint carefully and present your argument methodically so that your audience bury into your argument. There are some options that can be referred to.
- Overview, Body, Conclusion (In the case that your audience need clarity in your presentation because you think there is a risk of them getting lost in your complicated points).
- Anecdote, Content, Conclusion (In the case that your audience might be bored even before you start).
- Plan, Benefits, Anecdote (In the case that you are attempting to convince an audience to establish something new and are facing an audience that demands your presentation be short and to the point).

Other Useful Information

Do:
- Choose a single background for the entire presentation.
- Use simple, clean fonts.
- Use a font size that can be seen from the back of the room.
- Write in bulleted format and use consistent phrase structure in lists.
- Provide essential information only. Use key words to guide the reader/listener through the presentation.
- Use direct, concise language. Keep text to a minimum.

- Provide definitions when necessary.
- Use white space to set off text and/or visual components.
- Make sure each slide logically leads to the next.
- Use a heading for each slide.

Don't:
- Clutter the slide with graphics.
- Use complicated fonts.
- Add superfluous information.
- Put down every word you are going to say.
- Use images if they will distract.
- Use hard to read color combinations, like black on blue. Try to use high contrast combinations

P. S. This guide can be tailored according to the different PowerPoint presentations.

A Sample PowerPoint Presentation

Figure 4-3 to Figure 4-10 illustrate XP using PowerPoint Presentation.

Introduction to Extreme Programming (XP)

Source: ExtremeProgramming.org home

Figure 4-3   Page 1

What is XP

- Extreme Programming (XP) is actually a deliberate and disciplined approach to software development. It was based on observations of what made computer programming faster and what made it slower. About eight years old, it has already been proven at many companies of all different sizes and industries world wide.
- XP is an important new methodology for two reasons:
  - it is a re-examination of software development practices that have become standard operating procedures.
  - It is one of several new lightweight software methodologies created to reduce the cost of software.
- XP is successful because it emphasizes:
  - Customer involvement and satisfaction
  - Team work
- XP improves a software project in four essential ways:
  - Communication
  - Simplicity
  - Feedback
  - Courage

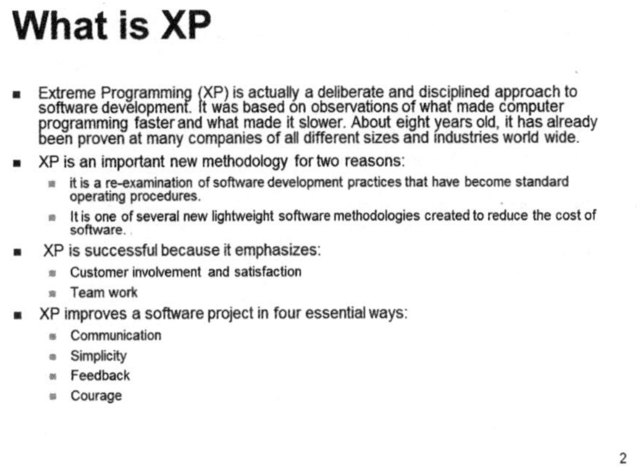

Figure 4-4   Page 2

When to use XP

- Dynamic requirements
- High project risks
- Small groups of programmers
- Testability
- Productivity

Figure 4-5　Page 3

The Rules and Practices of XP

- Planning
    - User stories are written.
    - Release planning creates the schedule.
    - Make frequent small releases.
    - The Project Velocity is measured.
    - The project is divided into iterations.
    - Iteration planning starts each iteration.
    - Move people around.
    - A stand-up meeting starts each day.
    - Fix XP when it breaks.
- Designing
    - Simplicity.
    - Choose a system metaphor.
    - Use CRC cards for design sessions.
    - Create spike solutions to reduce risk.
    - No functionality is added early.
    - Refactor whenever and wherever possible.

Figure 4-6　Page 4

The Rules and Practices of XP *(cont.)*

- Coding
    - The customer is always available.
    - Code must be written to agreed standards.
    - Code the unit test first.
    - All production code is pair programmed.
    - Only one pair integrates code at a time.
    - Integrate often.
    - Use collective code ownership.
    - Leave optimization till last.
    - No overtime.
- Testing
    - All code must have unit tests.
    - All code must pass all unit tests before it can be released.
    - When a bug is found tests are created.
    - Acceptance tests are run often and the score is published.

Figure 4-7　Page 5

XP Map

- The XP Map shows how they work together to form a development methodology. Unproductive activities have been trimmed to reduce costs and frustration.

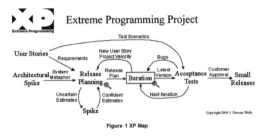

Figure 4-8　Page 6

What We Have Learned About XP

- Release Planning
- Simplicity
- System Metaphor
- Pair Programming
- Integrate Often
- Optimize Last
- Unit Tests
- Acceptance Tests

Figure 4-9　Page 7

More Information

- Web Sites
  - The Portland Pattern Repository
  - XProgramming.com
  - XP Developer
  - ...
- Books
  - ***Extreme Programming Explained: Embrace Change.*** By Kent Beck
  - ***Refactoring Improving the Design of Existing Code.*** By Martin Fowler
  - ***Extreme Programming Installed.*** By Ron Jeffries, Chet Hendrickson, and Ann Anderson
  - ....

Figure 4-10　Page 8

Exercises

1. Write a similar presentation regarding the Introduction to Software Requirement.

2. Write a similar presentation regarding the Introduction to Software Project Plan.

Unit 5

Designing the System

系统设计

Part 1

Listening & Speaking

Unit 5

Dialogue: Software Design

Sharon: Having completed the analysis modeling and requirements specification of the system, we can progress further!

Kevin: Yes, now let's have a discussion on our generic task set in the design process.

Jason: We will move from a "big picture" **view** of the software to a more narrow view that defines the details required to implement this system.[1] The process begins by focusing on architecture, I think.

[1] Replace with: We will perform the software design from a high level to more detailed levels step by step.

Sharon: What do you mean?

Jason: Using the analysis model we constructed last week, we need to select an architectural style that is appropriate for our software.

Kevin: After that, we can partition the analysis model into design subsystems and allocate these subsystems within the architecture.

Jason: In the detailed design phase, we must be certain that[2] each subsystem is functionally cohesive, and design subsystem interfaces first, and then allocate analysis classes or functions to each subsystem.

[2] Replace with:
1. We must ensure that…
2. We must make sure that…

Kevin: Additionally, it is an important factor for successful software to have a user interface **appreciated** by its users, particularly for business software like ours.

Jason: Sharon is a **competent** art designer.

Sharon: I'll do my best.

Kevin: We will collaborate to specify action sequences based on user scenarios and help Sharon to create a behavioral model of the interface.

Sharon: Ok, next, I will define interface objects and control mechanisms and beg you to review and revise as required.

Kevin: No problem. After completing the architecture design, Jason and I will conduct component-level design by specifying all algorithms at a relatively low level of abstraction, define component level data structure, and refine the interface of each component if necessary.

Jason: After reviewing the component level design, we can develop a deployment model for our system.

 **Exercises**

Work in a group, and make up a similar conversation by replacing the statements with other expressions on the right side.

 **Words**

view[vju:] n. 视图,观点 appreciate[ə'pri:ʃieit] v. 赏识,欣赏,重视	competent[ˈkɔmpitənt] adj. 有能力的,胜任的 beg[beg] v. 请求,恳求

Listening Comprehension: Software Design

Listen to the article and answer the following 3 questions based on it. After you hear a question, there will be a break of 15 seconds. During the break, you will decide which one is the best answer among the four choices marked (A), (B), (C) and (D).

Questions

1. Which description is not correct about the design process?
   (A) The design process moves from a macro-view of software to a more microcosmic view.

(B) Through the design process, requirements are translated into a "blueprint" for construction.

(C) The design process is represented from a high level of abstraction to lower levels step by step.

(D) It is reasonable for the design process to create a model of software with all details directly.

2. How many different elements are encompassed in the design model according to this article?
   (A) Two
   (B) Three
   (C) Four
   (D) Five

3. Which statement is not correct about the design concepts?
   (A) Design concepts have been developing over the first half-century of software engineering work.
   (B) The design concepts used in different projects may be inconsistent because of the different software engineering process that is chosen.
   (C) The design methods applied in the projects may be different in terms of the different software engineering process that is chosen.
   (D) Design concepts have evolved along with software engineering.

Words

commence[kəˈmens] v. 开始
sift[sift] v. 详审,精选
encompass[inˈkʌmpəs] v. 包含,包括

populate[ˈpɔpjuleit] v. 构成,填充
house[haus] v. 安置,存放

Phrases

come to a conclusion　得出结论,告一段落
converge on　集中于

Dictation: User Interface Design

This article will be played three times. Listen carefully, and fill in the numbered spaces with the appropriate words you have heard.

Unit 5 Designing the System 系统设计

When the concept of the interface first began to emerge, it was commonly understood as the hardware and software through which a human and a computer could communicate. So User Interface Design is also known as ___1___ Interaction or HCI. As it has evolved, the concept has come to include the **cognitive** and emotional aspects of the user's ___2___ as well.

Many technological innovations ___3___ upon User Interface Design to **elevate** their technical complexity to a ___4___ product. Technology alone may not win user ___5___ and subsequent **marketability**. The User Experience, or how the user experiences the end product, is the ___6___ to acceptance. And that is where User Interface Design enters the design ___7___. While product engineers ___8___ on the technology, usability ___9___ focus on the UI. For greatest efficiency and cost effectiveness, this working ___10___ should be maintained from the start of a project to its **rollout**.

UI affects the feelings, the emotions, and the mood of your users. If the UI is wrong and the user feels like they can't ___11___ your software, they **literally** will be angry and ___12___ it on your software. ___13___, if the UI is smart and things **work the way** the user expected them to work, they will be cheerful as they manage to ___14___ small goals. Thus, the **cardinal axiom** of all user interface design is that, a UI is ___15___ when the program ___16___ exactly how the user thought it would.

To make user happy, you have to ___17___ them in control of their environment. To do this, you need to correctly interpret their actions. Additionally, you should reduce the user's memory ___18___ and make the interface ___19___. Those three points above are considered as "___20___" and actually form the basis for a set of user interface design principles that guide this important software design action.

 Words

cognitive['kɔgnətiv] *adj*. 认知的，认识的	rollout['rəʊlaʊt] *n*. 首次展示,(产品或服务的)正式推出
elevate['eliveit] *v*. 提高,提升	literally['litərəli] *adv*. 真实地,确切地
marketability[ˌmɑːrkɪtə'bɪləti] *n*. 可销售性,可流通性	cardinal['kɑːdɪnl] *adj*. 主要的,最重要的
	axiom['æksɪəm] *n*. 公理,格言,规则

 Phrases

work the way 进行	

127

Part 2

Reading & Translating

Section A: Software Design

Design is defined in [IEEE 610.12—1990] as both "the process of defining the architecture, components, interfaces, and other characteristics of a system or component" and "the result of that process." Viewed as a process, software design is the software engineering life cycle activity in which software requirements are analyzed in order to produce a description of the software's internal structure that will **serve** as the basis for its construction.[1] More precisely, a software design (the result) must describe the software architecture — that is how software is decomposed and organized into components — and the interfaces between those components. It must also describe the components at a level of detail that enable their construction.[2]

Software design plays an important role in developing software: it allows software engineers to produce various models that form a kind of **blueprint** of the solution to be implemented (Figure 5-1). We can analyze and evaluate these models to determine whether or not they will allow us to fulfill the various requirements. We can also examine and evaluate various alternative solutions and **trade-offs**. Finally, we can use the resulting models to plan the subsequent development activities, in addition to using them as input and the starting point of construction and testing.

Figure 5-1 Software Design in Developing Software

In a standard listing of software life cycle processes such as IEEE/EIA 12207 Software Life Cycle Processes [IEEE 12207.0—1996], software design consists of two activities that fit between software requirements analysis and software construction:

- Software architectural design (sometimes called top level design): describing software's top-level structure and organization and identifying the various components.
- Software detailed design: describing each component sufficiently to allow for its construction.

General Strategies

Some often-cited examples of general strategies useful in the design process are divide-and-conquer and stepwise refinement, top-down vs. bottom-up strategies, data abstraction and information hiding, use of heuristics, use of patterns and pattern languages, use of an iterative and incremental approach.

Function-Oriented (Structured) Design

This is one of the classical methods of software design, where decomposition centers on identifying the major software functions and then elaborating and refining them in a top-down manner. Structured design is generally used after structured analysis, thus producing, among other things, data flow diagrams and associated process descriptions. Researchers have proposed various strategies (for example, transformation analysis, transaction analysis) and heuristics (for example, fan-in/fan-out, scope of effect vs. scope of control) to transform a DFD into a software architecture generally represented as a structure chart.

Object-Oriented Design

Numerous software design methods based on objects have been proposed. The field has evolved from the early object-based design of the mid-1980s (noun = object; verb = method; adjective = attribute) through OO design, where inheritance and polymorphism play a key role, to the field of component-based design, where meta-information can be defined and accessed (through reflection, for example). Although OO design's roots stem from the concept of data abstraction, responsibility-driven design has also been proposed as an alternative approach to OO design.

Data-Structure-Centered Design

Data-structure-centered design (for example, Jackson [3], Warnier-Orr [4]) starts from the data structures a program manipulates rather than from the function it performs. The software engineer first describes the input and output data structures (using Jackson's structure diagrams, for instance) and then develops the program's control structure based on these data structure diagrams. Various heuristics have been proposed to deal with special cases — for example, when there is a mismatch between the input and output structures.

Component-Based Design (CBD)

A software component is an independent unit, having well-defined interfaces and dependencies that can be composed and deployed independently. Component-based design addresses issues related to providing, developing, and integrating such components in order to improve reuse.

Other Methods

Other interesting but less mainstream approaches also exist: formal and rigorous methods and transformational methods.

Words

blueprint['blu:print] n. 蓝图,计划
trade-off 折中,平衡
often-cited 经常被引用的
divide-and-conquer 分而治之
stepwise['stepwaiz] adj. 逐步的
heuristics[hju'ristiks] n. 启发法,启发式教学法
iterative['itərətiv] adj. 迭代的,重复的,反复的
decomposition[ˌdi:kɔmpə'ziʃn] n. 分解
fan-in 扇入,输入端
fan-out 扇出,输出端
inheritance[in'herit(ə)ns] n. 继承
polymorphism[ˌpɔli'mɔ:fizm] n. 多态
meta-information 元信息
reflection[ri'flekʃ(ə)n] n. 反射
manipulate[mə'nipjuleit] v. 控制,操作,使用
well-defined 明确的
reuse[ri:'ju:z] n. 复用

Phrases

serve as 充当,用作
center on 集中在,着重在
among other things 其中
stem from 基于,处于
formal and rigorous methods 形式化和精确方法

Abbreviations

EIA Electronic Industries Association 美国电子工业协会
DFD Data Flow Diagram 数据流图
OO Object-Oriented 面向对象的

Notes

[1] **Original**: Viewed as a process, software design is the software engineering life

Unit 5　Designing the System　系统设计

cycle activity in which software requirements are analyzed in order to produce a description of the software's internal structure that will serve as the basis for its construction.

Translation：作为一个过程来看，软件设计是一项软件工程生命周期活动，其中，软件需求被分析以产生软件内部结构的描述，它将作为软件构建的基础。

［2］　**Original**：More precisely, a software design (the result) must describe the software architecture — that is how software is decomposed and organized into components — and the interfaces between those components. It must also describe the components at a level of detail that enable their construction.

Translation：更确切地讲，软件设计（的结果）必须描述软件体系结构——软件如何被分解和组织成组件——以及这些组件之间的接口。软件设计也必须在使其能够构建的详细级别上描述组件。

［3］　面向数据结构的 Jackson 方法的基本思想是：在充分理解问题的基础上，找出输入数据、输出数据的层次结构的对应关系，将数据结构的层次关系映射为软件控制层次结构，然后对问题的细节进行过程性描述。

［4］　面向数据结构的 Warnier-Orr 方法的基本思想是：使用顺序、选择、重复 3 种结构成分来表示数据的层次结构，进而导出程序的结构；使用 Warnier 图表示数据的层次结构。

　Exercises

Ⅰ. Read the following statements carefully, and decide whether they are true (T) or false (F) according to the text.

　　____ 1. In software design, software engineers form a blueprint of the solution to be implemented.

　　____ 2. Structure Chart is a kind of diagram for representing the software architecture in Function-Oriented Design.

　　____ 3. Object-Oriented Design has the same meaning but a different name with object-based design emerged in the mid-1980s.

　　____ 4. Inheritance and polymorphism are two of the most key characteristics in OO design.

　　____ 5. A software component has well-defined interfaces and dependencies that can be composed and deployed independently.

Ⅱ. Choose the best answer to each of the following questions according to the text.

1. Which of the following statements is wrong about the software design according to this article?

　　(A) Software design plays an important role in transition from software requirements to its construction.

　　(B) Software design describes the software architecture, interfaces and components, and defines other characteristics of a system or component.

(C) Using heuristics and an iterative approach as general strategies, software design is a heuristic and iterative process in nature.

(D) Software design allows software engineers to analyze software requirements and form a model of the requirements finally.

2. Which of the following statements is right about the software design approaches mentioned in this article?

(A) Structured design focuses on identifying the major software structure and then elaborating and refining them in a top-down manner.

(B) In Data-Structure-Centered Design, the software engineer first produces the input and output DFDs and then develops the program's control structure.

(C) In Object-Oriented Design, every noun can be strictly abstracted into an object, and every verb can be abstracted into a method as well as every adjective can be abstracted into an attribute of the object related.

(D) The main objective of Component-Based Design is improving software reuse.

3. How many kinds of methods for software design are referred to in this article in all?

(A) Five

(B) Six

(C) Seven

(D) Eight

Ⅲ. Fill in the numbered spaces with the words or phrases chosen from the box. Change the forms where necessary.

> give   build   depend   long   help
> compare   look   write   understand   follow

Applying Cohesion

  ___1___ on the type of software you are writing, you may need to compromise a bit. Although we should always strive to have our code at the highest level of cohesion, sometimes that may make the code ___2___ unnatural. There is a difference between being unaware of design principles and consciously not following a design principle in a ___3___ context. I don't write my code with the goal that it should satisfy every single design principle out there but I always try to have a good reason every time I decide not to ___4___ certain principles. Having said that, cohesion is one of the most important building blocks of software design and ___5___ it well is essential for writing well-crafted code.

If you are ___6___ a framework, a very generic part of your code, or data transformation, chances are that the majority of your modules and processing elements will be at sequential and functional levels. However, when ___7___ business rules in a commercial application, i.e. an application where there are business logic, user journeys, database access, etc., there is a good chance that some of your modules and processing elements will be at communicational level and some even at a lower level of cohesion. And that's OK as ___8___ as it was a conscious decision and the right thing to do in that context.

Some people ___9___ cohesion to the Single Responsibility Principle (SRP). Although SRP is a great software principle and entirely based on cohesion, it has a quite narrow and subjective scope.

Identifying responsibilities is not always an easy thing. We need to develop a keen eye to detect minor variations in behavior. Unit tests for a module can potentially ___10___ us to identify the different behaviors, if the code was really test-driven, of course.

The more cohesive your code is, the more reusable, robust and easy to maintain it will be.

Ⅳ. **Translate the following passage into Chinese.**
Design Reuse

Design reuse is the process of building new software applications and tools by reusing previously developed designs. New features and functionalities may be added by incorporating minor changes.

Design reuse involves the use of designed modules, such as logic and data, to build a new and improved product. The reusable components, including code segments, structures, plans and reports, minimize implementation time and are less expensive. This avoids reinventing existing software by using techniques already developed and to create and test the software.

Section B: What is the Difference between UX Design and UI Design?

UX and UI are interchangeably used so often that we think they mean the same thing. But there is a difference between these two terms: UX Design and UI Design (Figure 5-2). By understanding the difference between them, you will be able to build your product with the right design process. It is common for product managers, founders, CEOs, hiring managers to refer to design as "UX/UI design" when these are 2 dissimilar work scope. This can be seen by the sheer amount of job posts looking for UX/UI designers.

Figure 5-2　A Difference between UX Design and UI Design

　　User Experience (UX) Design is the process of researching, developing and improving all aspects of the users' interaction with the product in order to satisfy its users. The main goal is to improve the users' experience with the product, this is done through conducting usability tests, evaluating the results and refining the design—ultimately creating a product that is not only useful and valuable, but also easy and **pleasing** to use.

　　UX design is vital in contributing to a product that provides an effective and successful user experience, even **accommodating to** human error. It includes various disciplines such as interaction design, visual design, usability and more.

　　In a nutshell:
- User experience design is NOT about **visuals** or **mockups**, it focuses on the overall process/service flow, it's usability and experience. UX designers **craft** these experiences, thinking about how the experience makes the user feel, and how easy it is for the user to complete their desired task.
- User experience design is the process of developing and improving the quality of interaction between the user and all **facets** of the company.
- The ultimate purpose of UX design is to create usable, efficient, relevant, easy and **all-round** pleasant experiences for the user.

　　User Interface (UI) Design determines the visual appearance of a product, with the aim of visually designing the product for effective user experience. A good UI design attracts users and aids in providing a pleasant experience.

　　The elements of UI design include documents, images, videos, text, forms, buttons, **checkboxes**, **drop-downs** along with their respective interactions, the behavior of what will happen when the user interacts through clicking, dragging or **swiping**. The main goal is to create a **compelling** and beautiful user interface that **elicits** an emotional response and balance with good usability

　　The purpose of UI design is to make sure the user interface of a product is as **intuitive** as possible, and that means treating each and every visual and interactive element with careful consideration.

　　In a nutshell:
- User interface design conveys the brand's strengths and visual assets to a product's interface, ensuring the design is consistent, **coherent** and **aesthetically** pleasing.

- User interface design considers all the visual and interactive elements of a product's interface—including icons, spacing, buttons, color schemes, typography and responsive design.
- The ultimate goal of UI design is to visually guide the user through the product's interface, creating an intuitive experience that doesn't require the user to think too much.

UX design and UI design always overlap but there is a distinct difference.

In terms of functionality, UX is how things work, UI is how things look. UX is a process, whereas UI is a deliverable.

It's important to note that UX and UI go together hand-in-hand: you can't have one without the other (at least you shouldn't). A good way to understand the difference is to look into the different aspects of what UX designers and UI designers do.

A UX designer considers the user's entire journey throughout the product to accomplish their task: what steps does the user take? What tasks do they need to complete? Is the experience straightforward?

Much of a UX designer's work goes into investigating problems and pain-points that users come across and how to solve them. A UX designer, often working with a User Researcher, conducts extensive user research to first find out who the target users are, and what their needs are in relation to the product. After that, they will map out the user journey, adding in the information architecture—how the content is organized and labelled, and the various features the user needs. In the end, a wireframe is created, this is a blueprint or skeleton of the product.

With the blueprint, the UI designer steps in to bring it to life by considering all the visual aspects of the user journey. They consider all the individual screens and the respective touchpoints that the user may encounter—tapping a button, swiping through an image gallery or scrolling through a page. Asking questions like "How can different color schemes and combinations be used to create sufficient contrast and enhance readability?" or "How can we pair colors to cater to those who are color blind?".

UX design focuses on wireframing and prototyping, interaction design and user testing whereas UI design includes visual design and interaction design. It's important to note that both UX design and UI design require interaction design, which guides the user's feelings and behavior.

So here is a summary:
- UX design usually comes first in the product development process, followed by UI design.
- The UX designer maps out the bare bones of the user journey and the UI designer filled it in with visual and interactive elements.
- UX design is about identifying and solving user problems whereas UI design is about creating intuitive and aesthetically-pleasing interactive interfaces.

Words

sheer[ʃiə(r)] adj. 纯粹的,绝对的
post[pəust] n. 岗位
please[pli:z] v. 使喜欢,使高兴,使满意
visual['viʒuəl] n. 视觉资料(指说明性的图片、影片等)
mockup['mɔkʌp] n. 实物模型
craft[krɑ:ft] v. 精巧地制作
facet['fæsit] n. 面,方面
all-round 全面的,多方面的,综合性的
checkbox['tʃekbɔks] n. 复选框,检查框
drop-down 下拉菜单
swipe[swaip] v. 滑动屏幕
compelling[kəm'peliŋ] adj. 引人注目的,迷人的

elicit[i'lisit] v. 得出,诱出,引起
intuitive[in'tju:itiv] adj. 直觉的,凭直觉获知的
coherent[kəu'hiərənt] adj. 一致的,明了的,清晰的
aesthetical[i:s'θetikəl] adj. 美的,美学的
typography[tai'pɔgrəfi] n. 排版
hand-in-hand 手牵手的,亲密的,并进的
pain-point 痛点
wireframe['waiəfreim] n. 示意图,线框模型
pair[peə(r)] v. 把……组成一对

Phrases

accommodate to   适应
in a nutshell   概括地说
color scheme   配色方案,色彩设计
look into   调查,窥视,观察
map out   设计,规划,安排
step in   介入,插手干预
bring it to life   使……充满活力
cater to   迎合,为……服务
color blind   色盲
bare bone   基本框架,梗概

Abbreviations

CEO   Chief Executive Officer   首席执行官,执行总裁

Exercises

Ⅰ. Read the following statements carefully, and decide whether they are true(T) or false

（F）according to the text.

_____ 1. UI design usually comes first in the product development process, followed by UX design.

_____ 2. The UI designer maps out the bare bones of the user journey and the UX designer filled it in with visual and interactive elements.

_____ 3. UI design is about identifying and solving user problems whereas UX design is about creating intuitive and aesthetically-pleasing interactive interfaces.

_____ 4. Both UX design and UI design require interaction design, which guides the user's feelings and behavior.

_____ 5. UX and UI go together hand-in-hand: you can't have one without the other (at least you shouldn't).

Ⅱ. Choose the best answer to each of the following questions according to the text.

1. Which of the following statements is wrong about UX design?
    (A) User experience design is NOT about visuals or mockups, it focuses on the overall process/service flow, it's usability and experience.
    (B) User experience design is the process of developing and improving the quality of interaction between the user and all facets of the company.
    (C) The ultimate purpose of UX design is to create usable, efficient, relevant, easy and all-round pleasant experiences for the user.
    (D) The ultimate goal of UX design is to visually guide the user through the product's interface, creating an intuitive experience that doesn't require the user to think too much.

2. Which of the following statements is wrong about UI design?
    (A) User interface design conveys the brand's strengths and visual assets to a product's interface, ensuring the design is consistent, coherent and aesthetically pleasing.
    (B) User interface design considers all the visual and interactive elements of a product's interface—including icons, spacing, buttons, color schemes, typography and responsive design.
    (C) The ultimate goal of UI design is to visually guide the user through the product's interface, creating an intuitive experience that doesn't require the user to think too much.
    (D) The ultimate purpose of UI design is to create usable, efficient, relevant, easy and all-round pleasant experiences for the user.

3. Which of the following statements is right about UX design and UI design?
    (A) UI design usually comes first in the product development process, followed by UX design.
    (B) In terms of functionality, UX is how things work, UI is how things look.

(C) The UI designer maps out the bare bones of the user journey and the UX designer filled it in with visual and interactive elements.

(D) UI design is about identifying and solving user problems whereas UX design is about creating intuitive and aesthetically-pleasing interactive interfaces.

Ⅲ. **Fill in the numbered spaces with the words or phrases chosen from the box. Change the forms where necessary.**

> enable support depict short base
> order serve meet write use

Object Oriented Design

Object Oriented Design (OOD) ___1___ as part of the object oriented programming (OOP) process of lifestyle. It is mainly the process of ___2___ an object methodology to design a computing system or application. This technique ___3___ the implementation of a software ___4___ on the concepts of objects. Additionally, it is a concept that forces programmers to plan out their code in ___5___ to have a better flowing program.

The origins of Object Oriented Design (OOD) is debated, but the first languages that ___6___ it included Simula and SmallTalk. The term did not become popular until Grady Booch ___7___ the first paper titled Object-Oriented Design, in 1982. The chief objective of this type of software design is to define the classes and their relationships, which are needed to build a system that ___8___ the requirements contained in the Software Requirement Specifications.

Moreover, it is the discipline of defining the objects and their interactions to solve a problem that was identified and documented during the Object Oriented Analysis (OOA). In ___9___, Object Oriented Design (OOD) is a method of design encompassing the process of object oriented decomposition and a notation for ___10___ both logical and physical models of the system under design.

Ⅳ. **Translate the following passage into Chinese.**
Design Pattern

A design pattern is a repeatable solution to a software engineering problem. Unlike most program-specific solutions, design patterns are used in many programs. Design patterns are not considered finished product; rather, they are templates that can be applied to multiple situations and can be improved over time, making a very robust software engineering tool. Because development speed is increased when using a proven prototype, developers using design pattern templates can improve coding efficiency and final product readability.

Part 3

Simulated Writing: Software Design Specification

An important product of the design process is a set of documents that describe the system to be built. As we have seen, one part must tell the customers and users in natural language what the system will do; a second part uses technical terminology to describe the system's structure, data, and functions. Thus, the contents of the two parts may overlap, but the ways of expressing them may not.

The design documents should contain a section, called the design rationale, outlining the critical issues and trade-offs that were considered in generating the design. This guiding philosophy helps the customers and other developers understand how and why certain parts of the design fit together.

The design also contains descriptions of the components of the system. One section should address how the users interact with the system.

Usually, a set of diagrams or formal notations describe the overall organization and structure of the system, including all levels of abstraction.

If the system is distributed, the configuration in the design is detailed enough to show the topology of the network, how the network nodes will access one another, and the allocation of functions to the nodes. If the system requirements include constraints on timing, or if the nodes of the network must be synchronized, the design describes how the timing will work. Similarly, the design contains information about the control and routing of messages. It may also include prescriptions for the integrity of the network: making sure that the data are accurate or can be recovered after a failure.

If the customer requires it, elements of the design may address monitoring system performance. In addition, there may be a manual override of the system, and the design describes how it will work. Other sections of the design documents may address fault location and isolation, system reconfiguration, or special security measures.

Finally, the design is cross-referenced with the requirements to demonstrate how the design is derived from them. This correspondence forces us to check for completeness and consistency. In addition, such a cross-reference will make enhancements or modifications easier to track later. For example, if a requirement changes, the cross-reference points to the corresponding design changes needed.

Software Design Specification Template

1. Introduction

This section (1-2 pages) provides an overview of this entire document.

1.1 Project Overview

Describe the client, the problem to be solved, and the intended users. Explain the context in which your software will be used, i.e. the big picture. (1-3 paragraphs)

1.2 Project Scope

Mention the most important features of the system, inputs, data stores, and outputs. Do not discuss implementation details. Note any major constraints. (1-5 paragraphs)

1.3 Document Preview

Describe the purpose, scope of this document, and intended audience of this document. Mention the major sections that follow. Provide references to companion documents. (1-2 paragraphs)

2. Architectural Design

This section (2-4 pages) provides an overview and rationale for the program's data and architectural design decisions.

2.1 Section Overview

Provide a summary of the contents of this section. (1-2 paragraphs)

2.2 General Constraints

Describe global limitations or constraints that have a significant impact on your system design. Examples include hardware and software environments, interface requirements, external data representations, performance requirements, network requirements, etc. (1-3 paragraphs)

2.3 Data Design

Describe the structure of any databases, external files, and internal data structures. You may wish to include references to appendices containing E-R diagrams, data, or file formats. (1-3 paragraphs)

2.4 Program Structure

Describe the architectural model chosen and the major components. Include a pictorial representation (or reference to an appendix block or class diagram) of the major components. (1-4 paragraphs)

2.5 Alternatives Considered

Discuss the alternative architectural models considered and justify your choice for your architectural design. (1-4 paragraphs)

3. Detailed Design

This section represents the meat of your document. Be as detailed as time allows.

3.1 Section Overview

Provide a summary of the contents of this section. (1-2 paragraphs)

3.2　Component n Detail (include a sub-section for each component)

A structured description usually works. For example, if your components are classes you may wish to include the following subsections.

3.2.1　Description

3.2.2　Data Members (include type, visibility, and description)

3.2.3　Methods (include English or pseudocode descriptions for each one)

4. **User Interface Design**

4.1　Section Overview

Provide a summary of the contents of this section. (1-2 paragraphs)

4.2　Interface Design Rules

Describe and justify the conventions and standards used to design your interface. (1-2 paragraphs)

4.3　GUI Components

Note the GUI components or API's provided in the development environment that you plan on using. (1 paragraph + table)

4.4　Detailed Description

Provide a detailed description of the user interface including screen images. You may prefer to reference an appendix containing the screen snapshots. (1-4 pages)

5. **Conclusion**

Provide an ending to this document with a mention of implementation and testing strategies resulting from this design. (1-2 paragraphs)

6. **Appendices (a list of possibilities)**

6.1　Database Entity-Relationship Diagram

6.2　Architectural Design Block Diagram(s)

6.3　Class Diagram(s)

6.4　Class Sequence Diagram(s)

6.5　User Interface Screen Snapshots

P.S. This template can be tailored according to the different projects.

A Sample Software Design Specification

Software Design Specification For Web Publishing System

Version 1.0

Victor Lu (Team leader)
Jack Liang
Jason Wang
Mary Zeng
David Chang

April 22, 2020

Unit 5　Designing the System 系统设计

Table of Contents

1. **Introduction**
   1.1　Purpose
   1.2　Scope
   1.3　Glossary
   1.4　References
   1.5　Overview of Document
2. **Deployment Diagram**
3. **Architectural Design**
   3.1　On-Line Journal System
      3.1.1　On-Line Journal
      3.1.2　Author Search Form
      3.1.3　Category Search Form
      3.1.4　Keyword Search Form
      3.1.5　Article Table
      3.1.6　Author Table
      3.1.7　Category Table
   3.2　Article Manager
      3.2.1　People Manager Form
      3.2.2　Article Manager Form
      3.2.3　Publisher Form
      3.2.4　Historical Society Database Interface
      3.2.5　Author Table
      3.2.6　Reviewer Table
      3.2.7　ActiveArticle Table
      3.2.8　Author Relationship Table
      3.2.9　Reviewer Relationship Table
      3.2.10　Category Relationship Table
      3.2.11　Article Table
4. **Data Structure Design**
   4.1　On-Line Journal Database-Articles Table
      4.1.1　Article Table
      4.1.2　Author Table
      4.1.3　Category Table
   4.2　Article Manager Database
      4.2.1　Author Table
      4.2.2　Reviewer Table
      4.2.3　ActiveArticle Table
      4.2.4　Author Relationship Table
      4.2.5　Reviewer Relationship Table
      4.2.6　Category Relationship Table
5. **User Interface Design**
   5.1　On-Line Journal User Interface
   5.2　Article Manager User Interface
6. **Real-Time Design**
7. **Help System Design**
8. **Use Case Realizations**
   8.1a　Search Article-Author
   8.1b　Search Article-Category
   8.1c　Search Article-Keyword
   8.2　Communicate
   8.3　Add Author
   8.4　Add Reviewer
   8.5a　Update Person-Author
   8.5b　Update Person-Reviewer
   8.6a　Update Article Status-Delete Reviewer
   8.6b　Update Article Status-Add/Delete Category
   8.7　Enter Communication
   8.8　Assign Reviewer
   8.9　Check Status
   8.10　Send Communication
   8.11　Publish Article
   8.12　Remove Article

1. Introduction

1.1 Purpose

This document contains the complete design description of the Web Publishing System. This includes the architectural features of the system down through details of what operations each code module will perform and the database layout. It also shows how the use cases detailed in the SRS will be implemented in the system using this design.

The primary audiences of this document are the software developers.

1.2 Scope

This system has two parts: On-Line Journal, using standard client-server architecture with a database on the server, and Article Manager, a repository architecture using a database. This system does not interact with any external system, although the Editor will use E-mail function with attachment outside of the system.

《Since we identified Domain (Analysis) classes in the SRS, we are utilizing them in the design》

1.3 Glossary

Glossary is shown in Table 5-1.

Table 5-1 Glossary

Term	Definition
Active Article	The document that is tracked by the system; it is a narrative that is planned to be posted to the public website
Author	Person submitting an article to be reviewed. In case of multiple authors, this term refers to the principal author, with whom all communication is made
Editor	Person who receives articles, sends articles for review, and makes final judgments for publications
Historical Society Database	The existing membership database (also HS database)
Keyword	Word or phrase used in searching by keyword. Case is not significant
Member	A member of the Historical Society listed in the HS database
Reader	Anyone visiting the site to read articles
Review	A written recommendation about the appropriateness of an article for publication; may include suggestions for improvement
Reviewer	A person that examines an article and has the ability to recommend approval of the article for publication or to request that changes be made in the article
User	Reviewer or Author

1.4 References

Team leader Victor Lu, Jack Liang, Jason Wang, Mary Zeng, David Chang: Software Requirements Specification for Web Publishing System (Version 1.0), Beihang

University, 2020.

1.5 Overview of Document

- Chapter 2 is a Deployment Diagram that shows the physical nodes on which the system resides. This allows a clear explanation of where each design entity will reside. No design unit may straddle two nodes but must have components on each, which collaborate to accomplish the service.
- Chapter 3 is the Architectural Design. This is the heart of the document. It specifies the design entities that collaborate to perform the functionality of the system. Each of these entities has an Abstract Specification and an Interface that expresses the services that it provides to the rest of the system. In turn each design entity is expanded into a set of lower-level design units that collaborate to perform its services.
- Chapter 4 is the basic Data Structure Design, which for this project is a relational database. While it is separated out here for emphasis, it is really the lowest level of the Architectural Design.
- Chapter 5 is on User Interface Design and discusses the methodology chosen, why it was chosen and why it is expected to be effective.
- Chapter 6 describes the structure of the real-time system.
- Chapter 7 describes the structure of the Help System.
- Chapter 8 exhibits the Use Case Realizations. The implementation of each use case identified in the SRS is shown using the services provided by the design objects.

2. Deployment Diagram

Deployment Diagram is as shown in Figure 5-3.

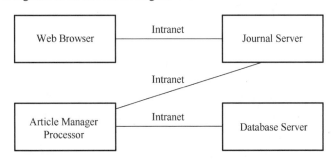

Figure 5-3 Deployment Diagram

A Reader accesses the On-Line Journal through the Internet using a Web browser (not part of this system although Web pages will run on it). The On-Line Journal resides on a dedicated Journal Server with a permanent Web connection. The Editor manages all of the article preparation work on his/her personal computer (the Article Manager Processor), communicating with the existing Historical Society Database on the Database Server when needed, and uploading completed articles to the Journal Server when they are ready. The Article Manager Processor contains a local file system and the Editor's E-mail system.

3. Architectural Design

Top-level Architectural Diagram is as shown in Figure 5-4.

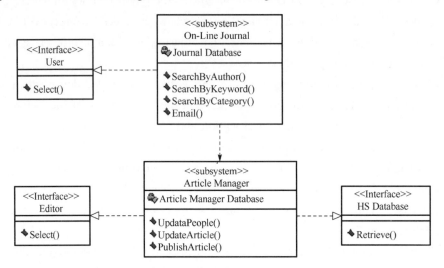

Figure 5-4 Top-level Architectural Diagram

User represents any of Author, Reader, or Reviewer. The User Interfaces are covered in Chapter 5.

《Please note that this top level design starts with the Domain Classes, which are subsystems in the design. Your design may not have such natural domain classes. In which case, your architectural design would look more like the design in Figure 5-5. Do not add domain classes unless there are natural ones for your project.》

3.1 On-Line Journal System

Name: On-Line Journal

Type: Subsystem

Node: Journal Server

Description: This is the primary entrance to the system for an Author, Reader or Reviewer. A reader can find and download articles from here. An author or reviewer can access their mail system from here.

Attributes: Journal Database (see section 4.1)

Resources: Client E-mail system

Operations (detailed below):

SearchByAuthor()

SearchByKeyword()

SearchByCategory()

E-mail()

Unit Design:

On-Line Journal Architectural Design is as shown in Figure 5-5.

Unit 5　Designing the System　系统设计

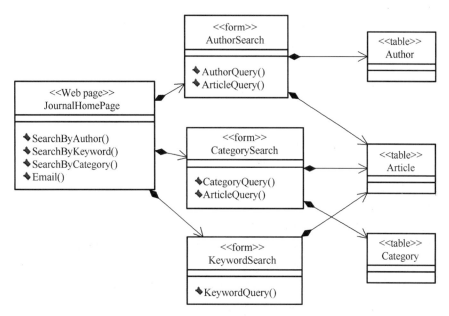

Figure 5-5　On-Line Journal Architectural Design

3.1.1　On-Line Journal

Name：On-Line Journal

Type：Web page（This unit has the same name as the subsystem and provides the operations listed there）

Node：Journal Server

Description：This class collects the functionalities of the interface of all users except the Editor. The code would be links on the home page of the website and would run on the client's browser.

Attributes：None

Resources：Client's system software

Operations：

(1)

Name：SearchByAuthor()

Arguments：None

Returns：File containing article

Description：The Reader will choose an author and an article by that author. An Abstract of the article will appear with the option to download.

Pre-condition：The Reader has had access to the correct Web page.

Post-condition：The article is downloaded to the Reader's PC.

Exceptions：The Reader may abandon the operation without downloading an article.

Flow of Events：

① The AuthorSearch form is linked to.

② This form calls AuthorQuery() which presents an alphabetical list of the authors in the database in a pull-down menu.

③ The Reader selects an author from the list.

④ This form calls ArticleQuery() for articles by that author.

⑤ The Reader selects an article.

⑥ The Abstract for that Article is displayed on the screen.

⑦ The Reader selects to download the article or to abandon the action.

(2)

Name：SearchByCategory()

Arguments：None

Returns：File containing article

Description：The Reader will choose a category and an article by that category. An Abstract of the article will appear with the option to download.

Pre-condition：The Reader has had access to the correct Web page.

Post-condition：The article is downloaded to the Reader's PC.

Exceptions：The Reader may abandon the operation without downloading an article.

Flow of Events：

① The CategorySearch form is linked to.

② This form calls CategoryQuery() which presents an alphabetical list of the categories in the database in a pull-down menu.

③ The Reader selects a category from the list.

④ This form calls ArticleQuery() for articles in that category.

⑤ The Reader selects an article.

⑥ The Abstract for that Article is displayed on the screen.

⑦ The Reader selects to download the article or to abandon the action.

(3)

Name：SearchByKeyword()

Arguments：None

Returns：File containing article

Description：The Reader will enter a keyword (or phrase) and choose an article with that keyword in its Abstract. An Abstract of the article will appear with the option to download.

Pre-condition：The Reader has had access to the correct Web page.

Post-condition：The article is downloaded to the Reader's PC.

Exceptions：The Reader may abandon the operation without downloading an article.

Flow of Events：

① The KeywordSearch form is linked to.

② This form presents a text box to enter a keyword.

③ This form calls KeywordQuery() which presents an alphabetical list of the titles of articles in the database which have that keyword in the article Abstract in a pull-down menu.

④ The Reader selects an article from the list.

⑤ The Reader selects to download the article or to abandon the action.

(4)

Name: E-mail()

Arguments: None

Returns: Access to user's mail system

Description: An author or Reviewer selects to pull up the E-mail system on the client PC with the address of the author inserted in the *To*: slot.

Pre-condition: The user has had access to the correct Web page.

Post-condition: The E-mail system with the editor inserted into the To: slot is pulled up.

Exceptions: The user may abandon the operation without accessing the E-mail system. If the system is not properly configured, this operation will fail.

Flow of Events:

The system invokes the *mailto*: tag in the HTML code on the Web page.

3.1.2 Author Search Form

Name: AuthorSearch

Type: Form

Node: Journal Server

Description: This form handles a search by author.

Attributes: Author Name, Article Title

Resources: None

Operations:

(1)

Name: AuthorQuery()

Arguments: None

Returns: Alphabetical list of authors

Description: This uses the query capability of the database to find an author.

Pre-condition: Control is on the AuthorSearch form.

Post-condition: An author name is returned.

Exceptions: May be abandoned at any time.

Flow of Events:

① The form upon loading calls a query which returns an alphabetical non-repeating list of authors in the database.

② The Reader selects an author.

(2)

Name: ArticleQuery()

Arguments: Author Name

Returns: Alphabetical list of titles of articles by that author

Description: This uses the query capability of the database to find the articles.

Pre-condition: Control is on the AuthorSearch form.

Post-condition: An article title is returned.

Exceptions: May be abandoned at any time.

Flow of Events:

① The form upon loading calls a query which returns an alphabetical list of articles by that author in the database.

② The Reader selects an article.

3.1.3 Category Search Form

Name: CategorySearch

Type: Form

Node: Journal Server

Description: This form handles a search by category.

Attributes: Category Name, Article Title

Resources: None

Operations:

(1)

Name: CategoryQuery()

Arguments: None

Returns: Alphabetical list of categories

Description: This uses the query capability of the database to find a category.

Pre-condition: Control is on the CategorySearch form.

Post-condition: A category name is returned.

Exceptions: May be abandoned at any time.

Flow of Events:

① The form upon loading calls a query which returns an alphabetical non-repeating list of category names in the database.

② The Reader selects a category.

(2)

Name: ArticleQuery()

Arguments: Category Name

Returns: Alphabetical list of titles of articles in that category

Description: This uses the query capability of the database to find the articles.

Pre-condition: Control is on the CategorySearch form.

Post-condition: An article title is returned.

Exceptions: May be abandoned at any time.

Flow of Events:

① The form upon loading calls a query which returns an alphabetical list of articles in that category in the database.

② The Reader selects an article.

3.1.4 Keyword Search Form

Name：KeywordSearch

Type：Form

Node：Journal Server

Description：This form handles a search by keyword.

Attributes：Article Name

Resources：None

Operations：

Name：KeywordQuery()

Arguments：None

Returns：Article Title

Description：This uses the query capability of the database to find articles.

Pre-condition：Control is on the KeywordSearch form.

Post-condition：An article title is returned.

Exceptions：May be abandoned at any time.

Flow of Events：

① The form presents a text box for the Reader to enter a keyword.

② The form calls a query which returns an alphabetical list of articles with that keyword in their Abstracts in the database.

③ The Reader selects an article.

3.1.5 Article Table

The Article Table is described in section 4.1.1.

3.1.6 Author Table

The Author Table is described in section 4.1.2.

3.1.7 Category Table

The Category Table is described in section 4.1.3.

3.2 Article Manager

Name：Article Manager

Type：Subsystem

Node：Article Manager Processor

Description：This is the primary entrance to the system for the Editor. All editorial duties can be performed here.

Attributes：Article Manager Database (see section 4.2)

Resources：Access to E-mail，the On-Line Journal database，and the Historical Society Database.

Operations (detailed below)：

UpdatePeople()

UpdateArticle()

PublishArticle()
Unit Design
Article Manager Architectural Design is as shown in Figure 5-6.

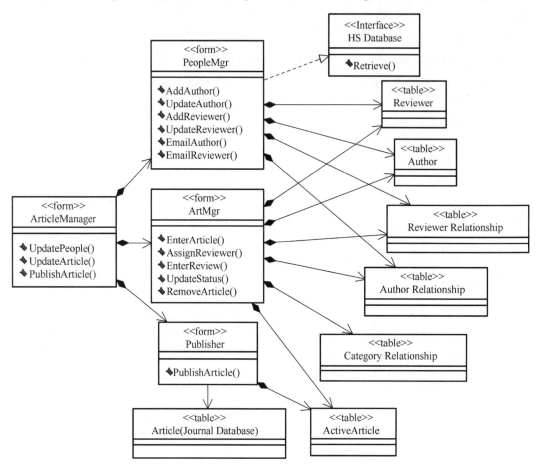

Figure 5-6 Article Manager Architectural Design

Name: Article Manager

Type: Form

Node: Article Manager Processor

Description: This allows the Editor to enter and update Authors, Articles, Reviewers, and to publish Articles.

Attributes: Article Manager Database (see section 4.2).

Resources: Access to E-mail, the On-Line Journal database, and the Historical Society Database.

Operations:

(1)

Name: UpdatePeople()

Arguments: None

Returns: Success/Failure

Description: This is used to enter or update authors and reviewers and to communicate with them.

Pre-condition: Form is active window.

Post-condition: See individual post-conditions.

Exceptions: May be abandoned at any time.

Flow of Events:

The Editor selects the function to be performed.

(2)

Name: UpdateArticle()

Arguments: None

Returns: Success/Failure

Description: This is used to enter or update any information about an article.

Pre-condition: Form is active window.

Post-condition: See individual post-conditions.

Exceptions: May be abandoned at any time.

Flow of Events:

The Editor selects the function to be performed.

(3)

Name: PublishArticle()

Arguments: None

Returns: Success/Failure

Description: This is used to move an article from the Article Manager to the On-Line Journal. It does not remove the article from the Article Manager.

Pre-condition: Form is active window.

Post-condition: Article has been copied to the On-Line Journal.

Exceptions: May be abandoned at any time.

Flow of Events:

The system calls the operation in Publisher.

3.2.1 People Manager Form

Name: PeopleMgr

Type: Form

Node: Article Manager Processor

Description: This is used to enter or update authors and reviewers and to communicate with them.

Attributes: Author Name, Reviewer Name

Resources: Access to Historical Society Database

Operations:

(1)

Name: AddAuthor()

Arguments: None.

Returns：Success/Failure

Description：This adds a new author to the Article Manager database.

Pre-condition：Form is active window and function has been selected.

Post-condition：Database has been modified.

Exceptions：May be abandoned at any time.

Flow of Events：

① The system presents a form to fill in Author information.

② The Editor fills in the form and submits it.

③ The system verifies that required fields have content and enters the new information into the database.

(2)

Name：UpdateAuthor()

Arguments：None

Returns：Success/Failure

Description：This updates author information in the Article Manager database.

Pre-condition：Form is active window and function has been selected.

Post-condition：Database has been modified.

Exceptions：May be abandoned at any time.

Flow of Events：

① The system presents an alphabetical drop-down list of Author's names using an SQL Server Query.

② The Editor selects an Author.

③ The system uses a query to retrieve the information about that author and displays it in a form to update Author information.

④ The Editor updates the form and submits it.

⑤ The system verifies that required fields have content and enters the updated information into the database.

(3)

Name：AddReviewer()

Arguments：None

Returns：Success/Failure

Description：This adds a new reviewer to the Article Manager database.

Pre-condition：Form is active window and function has been selected.

Post-condition：Database has been modified.

Exceptions：May be abandoned at any time.

Flow of Events：

① The system presents an alphabetical list of members' names from the Historical Society Database by executing an SQL Server Query.

② The Editor selects a name.

③ The system retrieves related information from the HS Database and presents a form to complete with Reviewer information.

④ The Editor fills in the form and submits it.

⑤ The system verifies that required fields have content and enters the new information into the database.

(4)

Name: UpdateReviewer()

Arguments: None

Returns: Success/Failure

Description: This updates reviewer information in the Article Manager database.

Pre-condition: Form is active window and function has been selected.

Post-condition: Database has been modified.

Exceptions: May be abandoned at any time.

Flow of Events:

① The system presents an alphabetical drop-down list of Reviewer's names using an SQL Server Query.

② The Editor selects a Reviewer.

③ The system uses a query to retrieve the information about that reviewer and displays it in a form to update Reviewer information.

④ The Editor updates the form and submits it.

⑤ The system verifies that required fields have content and enters the updated information into the database.

(5)

Name: E-mailAuthor()

Arguments: None

Returns: Access to user's mail system

Description: The Editor E-mails an Author.

Pre-condition: Form is active window and function has been selected.

Post-condition: The E-mail system with the author inserted into the To: slot is pulled up.

Exceptions: The user may abandon the operation without accessing the E-mail system. If the system is not properly configured, this operation will fail.

Flow of Events:

The system accesses the mail system with HTML tag mailto and fills in the To: slot on the E-mail form.

(6)

Name: E-mailReviewer()

Arguments: None

Returns: Access to user's mail system

Description: The Editor E-mails a Reviewer.

Pre-condition: Form is active window and function has been selected.

Post-condition: The E-mail system with the reviewer inserted into the *To*: slot is pulled up.

Exceptions: The user may abandon the operation without accessing the E-mail system. If the system is not properly configured, this operation will fail.

Flow of Events:

The system accesses the mail system with HTML tag mailto and fills in the To: slot on the E-mail form.

3.2.2 Article Manager Form

Name: ArtMgr

Type: Form

Node: Article Manager Processor

Description: This is used to enter a new article or to update all information about an article.

Attributes: Author Name, Article Name, Reviewer Name

Resources: None

Operations:

(1)

Name: EnterArticle()

Arguments: None

Returns: Success/Failure

Description: This adds a new article to the Article Manager database.

Pre-condition: Form is active window, function has been selected and all authors are already in the database.

Post-condition: Database has been modified.

Exceptions: May be abandoned at any time.

Flow of Events:

① The system presents a form to fill in Article information.

② The Editor uses a pull-down menu to assign author(s).

③ The Editor fills in the rest of the form and submits it.

④ The system verifies that required fields have content and enters the new information into the database.

(2)

Name: AssignReviewer()

Arguments: None

Returns: Success/Failure

Description: This adds a reviewer to an article in the Article Manager database.

Pre-condition: Form is active window, function has been selected and the article is already in the database.

Post-condition: Database has been modified.

Exceptions: May be abandoned at any time.

Unit 5　Designing the System　系统设计

Flow of Events:

① The system presents a pull-down menu of authors.

② The Editor selects an author.

③ The system presents a pull-down menu of articles by that author.

④ The Editor selects an article.

⑤ The system presents a pull-down menu of reviewers.

⑥ The Editor selects a reviewer.

⑦ The system enters the new information into the database.

(3)

Name: EnterReview()

Arguments: None

Returns: Success/Failure

Description: This adds a review to an article in the Article Manager database.

Pre-condition: Form is active window, function has been selected and the article with reviewer assigned is already in the database.

Post-condition: Database has been modified.

Exceptions: May be abandoned at any time.

Flow of Events:

① The system presents a pull-down menu of authors.

② The Editor selects an author.

③ The system presents a pull-down menu of articles by that author.

④ The Editor selects an article.

⑤ The system presents a pull-down menu of reviewers for that article.

⑥ The Editor selects a reviewer and enters the review.

⑦ The system enters the new information into the database including changing the status of the review pending.

(4)

Name: UpdateStatus()

Arguments: None

Returns: Success/Failure

Description: This allows the editor to review and update the status of any article in the Article Manager database.

Pre-condition: Form is active window and function has been selected.

Post-condition: Database has been modified.

Exceptions: May be abandoned at any time.

Flow of Events:

① The system presents a pull-down menu of authors.

② The Editor selects an author.

③ The system presents a pull-down menu of articles by that author.

④ The Editor selects an article.

⑤ The system presents a form with all the information about the article. Appropriate fields will have pull-down menus.

⑥ The Editor modifies selected fields.

⑦ The system enters the new information into the database.

(5)

Name：RemoveArticle()

Arguments：None

Returns：Success/Failure

Description：This removes an article in the Article Manager database.

Pre-condition：Form is active window, function has been selected and the article is already in the database.

Post-condition：Database has been modified.

Exceptions：May be abandoned at any time.

Flow of Events：

① The system presents a pull-down menu of authors.

② The Editor selects an author.

③ The system presents a pull-down menu of articles by that author.

④ The Editor selects an article.

⑤ The system prompts for confirmation of deletion.

⑥ The Editor confirms the deletion.

⑦ The system removes the article entry from the database.

3.2.3 Publisher Form

Name：Publisher

Type：Form

Node：Article Manager Processor

Description：This transfers an article from the Article Manager Database to the On-Line Journal Database.

Attributes：None

Resources：None

Operations：

Name：PublishArticle()

Arguments：None

Returns：Success/Failure

Description：This transfers an article from the Article Manager Database to the On-Line Journal Database.

Pre-condition：Form is active window and function has been selected.

Post-condition：Article is added to On-Line Journal Database.

Exceptions：May be abandoned at any time.

Flow of Events:

① The system presents a pull-down menu of authors.

② The Editor selects an author.

③ The system presents a pull-down menu of articles by that author.

④ The Editor selects an article.

⑤ The system presents a form with the article information.

⑥ The system accesses the Articles Table of the On-Line Journal Database and downloads a form for a new entry into that table.

⑦ The system transfers information from the article information form from the Article Manager Database into the article form from the On-Line Journal Database and asks for editing and/or confirmation.

⑧ The Editor modifies the entry, if desired, and confirms the entry.

3.2.4 Historical Society Database Interface

Name: Historical Society Database Interface

Type: Interface

Node: Article Manager Processor

Description: Provides membership information about a potential reviewer.

Attributes: None

Resources: Uses of Historical Society Database

Operations:

Name: Retrieve()

Arguments: None

Returns: Selected membership information

Description: Allows the Editor to search the HS Database for information about a reviewer, who must be a society member.

Pre-condition: Secure access has been established between this system and the HS Database System and an Add Reviewer Form is open.

Post-condition: Information is returned.

Exceptions: May be abandoned at any time.

Flow of Events:

① The Interface accesses the HS Database and retrieves an alphabetical list of all members and displays it as a pull-down menu.

② The Editor chooses an entry from the list.

③ The Interface retrieves selected information from the Database and inserts it into the open Reviewer form.

3.2.5 Author Table

The Author Table is described in section 4.2.1.

3.2.6 Reviewer Table

The Reviewer Table is described in section 4.2.2.

3.2.7 ActiveArticle Table

The ActiveArticle Table is described in section 4.2.3.

3.2.8 Author Relationship Table

The Author Relationship Table is described in section 4.2.4.

3.2.9 Reviewer Relationship Table

The Reviewer Relationship Table is described in section 4.2.5.

3.2.10 Category Relationship Table

The Category Relationship Table is described in section 4.2.6.

3.2.11 Article Table

The Article Table is described in section 4.1.1.

4. Data Structure Design

There are two databases embedded in this product. The first is the On-Line Journal Database and resides on the Journal Server. The second is the Article Manager Database and resides on the Editor's LAN.

4.1 On-Line Journal Database-Articles Table

This database contains three tables. Separating the Author and Article tables out allows unique storage for the basic article with multiple authors and/or categories supported. The SQL Server Query feature of the database management system is utilized extensively by the program.

4.1.1 Article Table

Article table is shown as follows (Table 5-2).

Table 5-2 Article Table

Field	Type	Description
ArticleID	Integer	Primary Key
Title	Varchar (80)	Title of Article
Abstract	Text	Used for Keyword Search, includes Title of Article
Content	Text	Body of Article
Article Address	Text	The File Address of the Article
Status	Bit	The Judgment of the Status of Article (Active or not)

4.1.2 Author Table

Author table is shown as follows (Table 5-3).

Table 5-3　Author Table

Field	Type	Description
ArticleID	Integer	Foreign
AuthorID	Integer	Foreign

This table has a primary key < ArticleID，AuthorID >. It can be searched for multiple authors for an article.

4.1.3 Category Table

Category table is shown as follows (Table 5-4).

Table 5-4　Category Table

Field	Type	Description
ArticleID	Integer	Foreign
Category	Varchar (20)	Topic of article

This table has a primary key < ArticleID，Category >. It can be searched for multiple categories for an article.

4.2 Article Manager Database

This database contains six tables. Separating the Relationship Tables out allowed for unique storage of Articles.

4.2.1 Author Table

Author table is shown as follows (Table 5-5).

Table 5-5　Author Table

Field	Type	Description
AuthorID	Integer	Primary Key
Author-Last	Varchar (20)	Last name of an author
Author-First	Varchar (50)	First name of an author
E-mail	Varchar (100)	E-mail address

4.2.2 Reviewer Table

Reviewer table is shown as follows (Table 5-6).

Table 5-6　Reviewer Table

Field	Type	Description
ReviewerID	Integer	Primary Key
Reviewer-Last	Varchar (20)	Last name of a reviewer
Reviewer-First	Varchar (50)	First name of a reviewer
E-mail	Varchar (100)	E-mail Address

4.2.3 Active Article Table

Active Article table is shown as follows (Table 5-7).

Table 5-7　Active Article Table

Field	Type	Description
ArticleID	Integer	Primary Key
Title	Varchar (80)	Title of Article
Abstract	Text	Includes Title of Article
Content	Text	Body of Article

4.2.4 Author Relationship Table

Author Relationship table is shown as follows (Table 5-8).

Table 5-8　Author Relationship Table

Field	Type	Description
ArticleID	Integer	Foreign
AuthorID	Integer	Foreign

This table has a primary key < ArticleID, AuthorID >. It can be searched for multiple authors for an article.

4.2.5 Reviewer Relationship Table

Reviewer Relationship table is shown as follows (Table 5-9).

Table 5-9　Reviewer Relationship Table

Field	Type	Description
id	Integer	Default increment primary key
ArticleID	Integer	Foreign
ReviewerID	Integer	Foreign
DateSent	Date	If not yet returned
Review	Text	When returned
CreateTime	Date	The time this row is added to database
Canceled	Bit	Review is removed from Reviewer

This table has a default increment primary key named id. It can be searched to see how many reviews are not yet returned for an article, how many reviews a reviewer has completed, and how many reviews not yet completed (with the date sent to the reviewer).

4.2.6 Category Relationship Table

Category Relationship table is shown as follows (Table 5-10).

Table 5-10 Category Relationship Table

Field	Type	Description
ArticleID	Integer	Foreign
Category	Text	Topic of Article

This table has a primary key < ArticleID，Category >. It can be searched for multiple categories for an article.

5. User Interface Design

5.1 On-Line Journal User Interface

The On-Line User interface will feature the logo of the Historical Society as background. There will be a welcoming message with instructions at the top，three large buttons in the middle，and a small button on the bottom right.

The text of the welcoming message and the instructions has not yet been approved by the Historical Society.

The three buttons will be labeled (from left to right) "Search by Author"，"Search by Category"，and "Search by Keyword". The small button will be labeled "E-mail Editor".

When either the Author or Category search is chosen，a pop-up window with a pull-down menu containing an alphabetical，non-repeating list of the appropriate names from the database will appear. When a name is chosen，the pop-up window will be replaced with another pop-up window containing the chosen name at top and all the article titles meeting that criterion.

As each title is selected，the Abstract will appear in the bottom of the form in a scrollable text box. There will be a button to "Select Article". Once an article is selected，the Web page will invoke a link HTML tag to present the article to open or save.

As authors and categories will be added to the database only when an article referencing them is added，there can be no extra authors or categories. Therefore，no error messages are needed for this part of the page.

The "E-mail Editor" button will invoke a mailto: HTML tag. This tag will invoke the client E-mail server with the To: slot of the E-mail header filled in. Any error condition will result in a message from the utility software on the client machine.

5.2 Article Manager User Interface

The UI of Article Manager is a form where users gain access to all article services (usually related to database manipulation). It will contain a list of all services grouped by type. The types are "Update People"，"Update Article" and "Publish Article".

The details of the various subsequent screens are functionally described in the rest of this document. The Editor has not expressed a desire for any interface beyond a strictly functional one. As most of the choices are from pull-down menus，there are few opportunities for error messages. The help system can be invoked directly from each page by a small button of the top right corner.

6. Real-Time Design

Real-time considerations are minimal in this project.

The Editor is the sole user of the Article Manager part of the system.

The On-Line Journal is designed for multiple users but the concurrent usage is handled client-side. That is, the Home Page will execute on the client (user's) computer and will make requests of the On-Line Journal server. These requests are handled sequentially by the server with no transient data storage.

7. Help System Design

There is only a minimal Help System for this project.

The On-Line Journal will have a button to E-mail Editor for a user to report problems.

The Article Manager will have a Help choice that will contain the following information:

a. An overview of the Editor's functionalities extracted from the SRS; and

b. A list of error messages and details on why they were generated.

In addition, the Help System will contain instructions on how the Editor can bypass the user interface and access the database directly.

8. Use Case Realizations

8.1a Search Article-Author

The sequential diagram for Search Article by Author is shown in Figure 5-7.

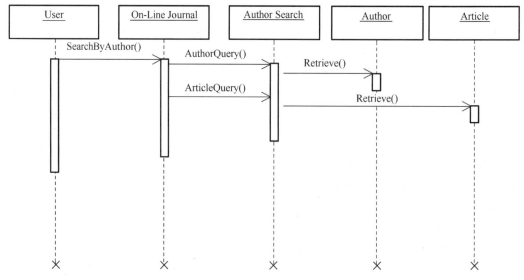

Figure 5-7 Sequential Diagram for Search Article by Author

Xref：SRS 3.2.1

8.1b Search Article-Category

The sequential diagram for Search Article by Category is shown in Figure 5-8.

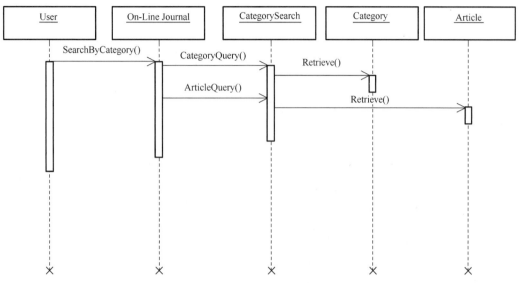

Figure 5-8 Sequential Diagram for Search Article by Category

Xref：SRS 3.2.1

8.1c Search Article-Keyword

The sequential diagram for Search Article by Keyword is shown is Figure 5-9.

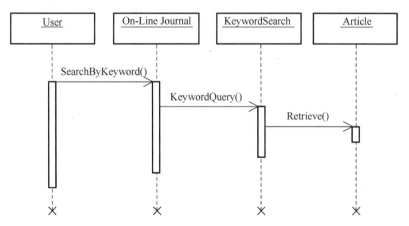

Figure 5-9 Sequential Diagram for Search Article by Keyword

Xref：SRS 3.2.1

8.2 Communicate

The sequential diagram for Communicate is shown in Figure 5-10. Remember：this system only helps to activate users' mail systems，and how their mail systems will work is beyond this system's scope.

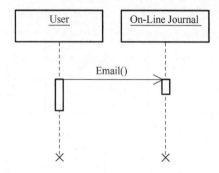

Figure 5-10　Sequential Diagram for Communicate

Xref：SRS 3.2.2

8.3　Add Author

The sequential diagram for Add Author is shown in Figure 5-11.

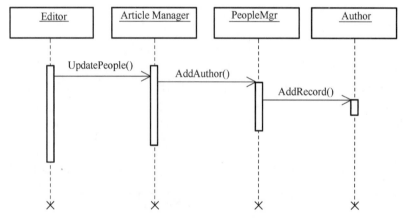

Figure 5-11　Sequential Diagram for Add Author

Xref：SRS 3.2.3

8.4　Add Reviewer

The sequential diagram for Add Reviewer is shown in Figure 5-12.

Xref：SRS 3.2.4

8.5a　Update Person-Author

The sequential diagram for Update Person by Author is shown in Figure 5-13.

Xref：SRS 3.2.5

8.5b　Update Person-Reviewer

The sequential diagram for Update Person by Reviewer is shown in Figure 5-14.

Xref：SRS 3.2.5

8.6a　Update Article Status-Delete Reviewer

The sequential diagram for Update Article Status by Delete Reviewer is shown in Figure 5-15.

Unit 5　Designing the System　系统设计

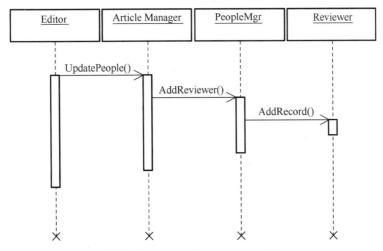

Figure 5-12　Sequential Diagram for Add Reviewer

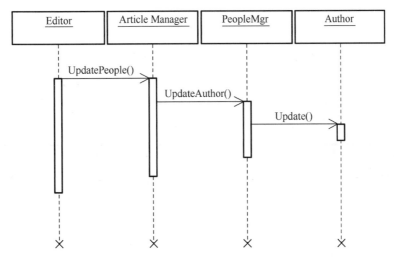

Figure 5-13　Sequential Diagram for Update Person by Author

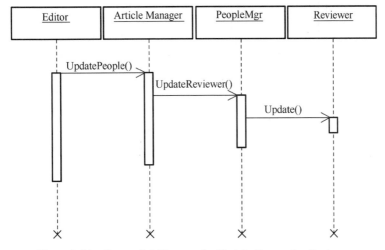

Figure 5-14　Sequential Diagram for Update Person by Reviewer

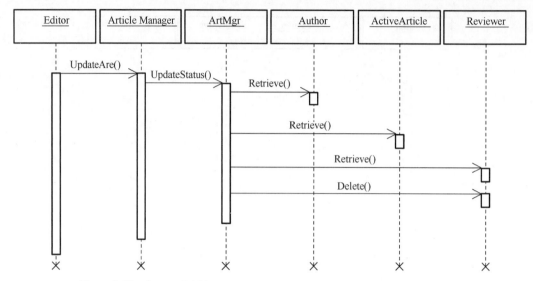

Figure 5-15   Sequential Diagram for Update Article Status by Delete Reviewer

Xref: SRS 3.2.6

《There actually should be more of these realizations for Use Case 3.2.6》

8.6b   Update Article Status-Add/Delete Category

The sequential diagram for Update Article Status by Add/Delete Category is shown in Figure 5-16.

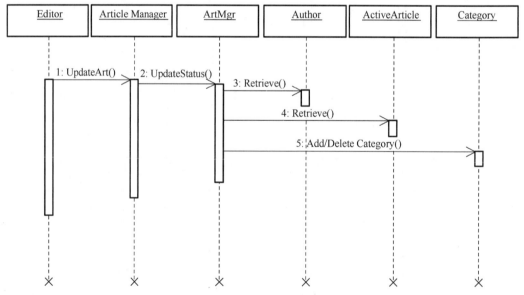

Figure 5-16   Sequential Diagram for Update Article Status by Add/Delete Category

Xref: SRS 3.2.6

8.7   Enter Communication

In the above, when the article is retrieved, modify its information and then update the record in the table with this modified information.

Xref: SRS 3.2.7

8.8 Assign Reviewer

The sequential diagram for Assign Reviewer is shown is Figure 5-17.

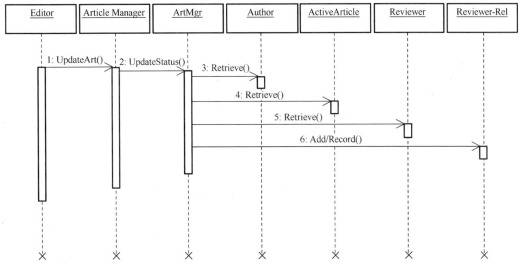

Figure 5-17 Sequential Diagram for Assign Reviewer

8.9 Check Status

Use section 8.6 above but make no changes.

Xref: SRS 3.2.9

8.10 Send Communication

The sequential diagram for Send Communication is shown in Figure 5-18.

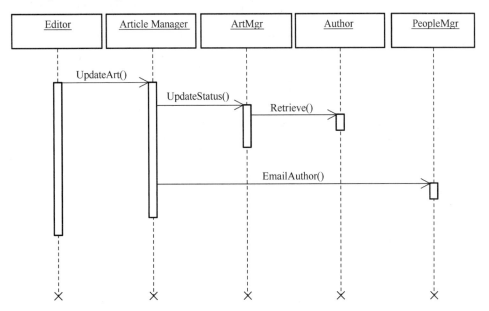

Figure 5-18 Sequential Diagram for Send Communication

Xref：SRS 3.2.10

8.11　Publish Article

The sequential diagram for Publish Article is shown in Figure 5-19.

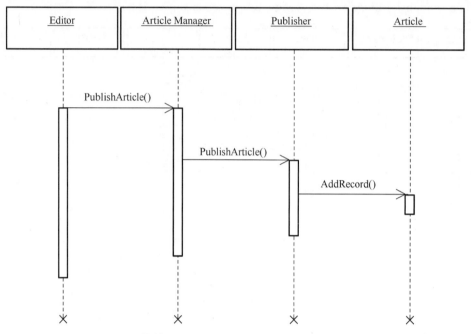

Figure 5-19　Sequential Diagram for Publish Article

Xref：SRS 3.2.11

8.12　Remove Article

The sequential diagram for Remove Article is shown is Figure 5-20.

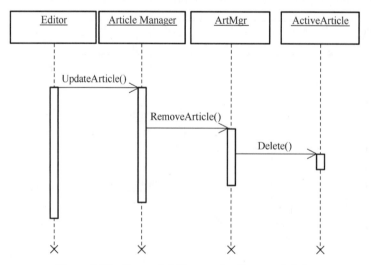

Figure 5-20　Sequential Diagram for Remove Article

Xref：SRS 3.2.12

 Exercises

1. What does this document mainly describe?
2. How many functions does Web Publishing System provide?
3. What is Article Manager Database mainly used for?

Unit 6

Implementing the System

系统实现

Unit 6　Implementing the System　系统实现

Part 1

Listening & Speaking

Unit 6

Dialogue：Creating High-Quality Code

Kevin： Before coding, I think it is necessary to **conform to** some principles during coding. **With the help of** these guidelines, we can be more reliable to create high-quality code and use the advantages of C♯ sufficiently, a popular object-oriented language which we have decided to use for coding.

Jason： For Object-Oriented Programming[1], programmers should think about programming in terms of classes, I think. A class is a collection of data and **routines** that share a cohesive, well-defined responsibility, and might also be a collection of routines that provides a cohesive set of services even if no common data is involved. A key to being an effective programmer is maximizing the portion of a program that you can safely ignore while working on any one section of code. Classes are the primary tool for accomplishing that objective.

Sharon： I absolutely agree with you. Furthermore, I think that the first and probably the most important step in creating a high-quality class is creating a good interface. This consists of creating a good abstraction for the interface to represent and ensure that the details remain hidden behind the abstraction. Particularly, in the analysis and design phase of our system, we have already identified all the classes, and defined the attributes and operation relevant for each of them. Now we should review and modify the classes with these criteria, which should be held constant when coding all of them.

[1] Replace with：
1. Considering Object-Oriented Programming
2. **With an eye to** Object-Oriented Programming
3. **With a view to** Object-Oriented Programming

Kevin: Yes. In addition[2], another important factor is high-quality routines within the classes. The routine makes programs easier to read and understand than any other feature of any programming language, and is also the greatest technique ever invented for saving space and improving performance, I think. Imagine how much larger our code would be if we had to repeat the code for every call to a routine instead of **branching to** the routine. Imagine how harder it would be to make performance improvements in the same code used in a dozen places instead of making them all in one routine. Once we've identified the class's major routines in the first step, we must construct each specific routine. Construction of each routine typically **unearths** the need for additional routines, both **minor** and major, and issues **arising from** creating those additional routines often **ripple back to** the overall class design.

> [2] Replace with:
> 1. Besides
> 2. Furthermore
> 3. Moreover

Jason: So constructing classes and constructing routines tends to be an iterative process.

Kevin: That's right. In every iterative process, we will test each routine as it's created. After the class as a whole becomes operational, we will review and test the class as a whole for any issues that can't be tested at the individual-routine level. And at each step, we should check our work and encourage others to check it too. In that way, we'll catch mistakes at the least expensive level, when we've invested the least amount of effort.

Sharon: Besides, the **Pseudocode** Programming Process is a useful tool for detailed design and makes coding easy. Pseudocode can be translated directly into **comments**, ensuring that the comments are accurate and useful.

Unit 6　Implementing the System　系统实现

 Exercises

Work in a group, and make up a similar conversation by replacing the statements with other expressions on the right side.

 Words

routine[ruːˈtiːn] n. 程序,例行程序
unearth[ʌnˈɜːθ] v. 发现,揭露
minor[ˈmainə(r)] adj. 较小的,次要的
pseudocode[ˈsjuːdəʊˌkəʊd] n. 伪代码
comment[ˈkɔment] n. 注释,注解

 Phrases

conform to　遵守
with the help of　借助
with an eye to　着眼于
with a view to　为了,目的在于
branch to　分支到
arise from　由……引起,起因于
ripple back to　回溯到

Listening Comprehension: Writing the Code

Listen to the article and answer the following 3 questions based on it. After you hear a question, there will be a break of 15 seconds. During the break, you will decide which one is the best answer among the four choices marked (A), (B), (C) and (D).

Questions

1. Which point of view is not correct on the purpose of matching implementation to design?
   (A) Easy to implement all the structures and relationships that are described with the charts and tables in the detailed design document as code directly
   (B) Easy to trace the algorithms, functions, interfaces, and data structures from design to code and back again
   (C) Easy to integrate all of the programmers' code into the whole system
   (D) Easy for testing, maintenance, and configuration management over time

2. Which advantage of the design characteristics is not referred in this article?
   (A) Low coupling

175

(B) Well-defined interfaces
(C) High cohesion
(D) Hierarchical building

3. Which technique is not mentioned in this article on coding with your team in mind?
(A) Standards
(B) Documents
(C) Common design techniques and strategies
(D) Information hiding

 Words

daunt[dɔːnt] v. 沮丧,使气馁
construct[kənˈstrʌkt] n. 结构体
correspondence[ˌkɔrəˈspɔndəns] n. 通信
modularity[ˌmɔdjuˈlæriti] n. 模块性
couple[ˈkʌpl] v. 耦合

cohesion[kəuˈhiːʒn] n. 内聚
algorithm[ˈælɡəriðəm] n. 算法
function[ˈfʌŋkʃn] n. 函数
invoke[inˈvəuk] v. 调用

 Phrases

fit in with   适应,符合,与……一致
take advantage of   利用
with ... in mind   把……放在心上,以……为目的

Dictation: Concentrate on the Vital Few, Not the Trivial Many

This article will be played three times. Listen carefully, and fill in the numbered spaces with the appropriate words you have heard.

The Pareto principle, also known as the 80-20 rule, states that, for many events, 80 percent of the ___1___ comes from 20 percent of the ___2___. In other words, you can get 80 percent of the result with 20 percent of the effort. It is a statistical method to identify the ___3___ of **agents** that **exert** the ___4___ effect.

Business management **thinker** Joseph M. Juran, father of Pareto principle in ___5___ management, suggested the principle and ___6___ it after Italian ___7___ Vilfredo Pareto, who observed that 80 percent of income in Italy went to 20 percent of the ___8___. It is a common **rule of thumb** which **applies to** a lot of areas. e.g., "80

percent of your sales comes from 20 percent of your clients" in business.
_____9_____, there are many software phenomena _____10_____ Pareto distribution. For example, 20 percent modules consume 80 percent resources, 20 percent modules _____11_____ 80 percent errors, 20 percent errors consume 80 percent repair costs, 20 percent modules consume 80 percent _____12_____ time, and 20 percent tools _____13_____ 80 percent usage, and so on.

One of definite applications of the software phenomena above is _____14_____ optimization. Donald Knuth **profiled** his line-count program and found that it was spending half its execution time in two _____15_____. He changed a few lines of code and _____16_____ the speed of the **profiler** in less than an hour. It is advised that you should measure the code to find the _____17_____ and then put your _____18_____ into optimizing the few _____19_____ that are used the most.

The Pareto Principle can and should be used by every intelligent person in their _____20_____ life. It can **multiply** your profitability and effectiveness.

Words

vital[ˈvaitl] *adj.* 至关重要的,生死攸关的	thinker[ˈθiŋkə(r)] *n.* 思想者
trivial[ˈtriviəl] *adj.* 不重要的,价值不高的,微不足道的	profile[ˈprəufail] *v.* 描……的轮廓,扼要描述
agent[ˈeidʒənt] *n.* 使然力(指引起一定作用的人或其他因素)	profiler[ˈprəufailə(r)] *n.* 曲线图,概览图,分析工具
exert[igˈzɜːt] *v.* 施以影响,施加(压力等)	multiply[ˈmʌltiplai] *v.* 使增加

Phrases

rule of thumb　单凭经验的方法
apply to　应用

Part 2

Reading & Translating

Section A: Computer Programming

A program is a list of **instructions** that the computer must follow to process data into information. The instructions consist of **statements** used in a programming language, such as Python. Examples are programs that do word processing, desktop publishing, or payroll processing. Programming is a five-step process for creating that lists of instructions as follows:

1. The problem clarification (definition) step consists of six mini-steps—clarifying program objectives and users, outputs, inputs, and processing tasks; studying the feasibility of the program; and documenting the analysis.

2. To design the solution, one first needs to create an algorithm. An algorithm is a formula or sets of steps for solving a particular problem. In the program design step, the software is designed in two mini-steps. First, the program logic is determined through a top-down approach and modularization, using a hierarchy chart (or structure chart). Then it is designed **in detail**, either in narrative form, using pseudocode, or graphically, using **flowcharts**. Pseudocode is a method of designing a program using normal human-language statements to describe the logic and the processing flow. Most programmers use a design approach called structured programming. Structured programming takes a top-down approach that breaks programs into modular forms. In structured program design, three control structures are used to form the logic of a program: sequence, selection, and iteration (or loop).

3. Once the design has been developed, the actual writing of the program begins. Writing the program is called **coding**. Coding is what many people think of when they think of programming, although it is only one of the five steps. Coding consists of translating the logic requirements from pseudocode or flowcharts into a programming language — the **letters**, numbers, and symbols that make up the program. A programming language is a set of rules that tells the computer what operations to do. Examples of well-known programming languages are Python, C, C++, and Java. These are called high-level languages.

4. Program testing involves running various tests and then running real-world data to make sure the program works. Two principal activities are **desk-checking** and debugging. These steps are called **alpha testing**. Desk-checking is simply **reading through**, or checking, the program to make sure that it's **free of** errors and that the logic works. Once the program has been desk-checked, further errors, or "**bugs**" will **doubtless surface**. To debug means to detect, **locate**, and remove all errors in a computer program. Mistakes may be syntax errors or logical errors. Syntax errors are caused by **typographical** errors and incorrect use of the programming language. Logic errors are caused by incorrect use of control structures. After desk-checking and debugging, the program may run **fine**—in the laboratory. However, it needs to be tested with real data; this is called **beta testing**.

5. Writing the program documentation is the fifth step in programming. The resulting documentation consists of written descriptions of what a program is and how to use it. Documentation is not just an **end-stage** process of programming. It has been (or

Unit 6 Implementing the System 系统实现

should have been) going on throughout all programming steps. Documentation consists of user documentation, operator documentation, programmer documentation, and it can be used to maintain the program.

Programming languages have five generations. The births of the generations are as follows:

(1) First generation, 1945—machine language

(2) Second generation, mid-1950s—assembly language

(3) Third generation, mid-1950s to early 1960s—high-level languages (procedural languages and object-oriented); for example, Fortran, COBOL, BASIC, C, and C++

(4) Fourth generation, early 1970s—very high-high-level languages (problem-oriented languages); for example, SQL, Intellect, NOMAD, FOCUS

(5) Fifth generation, early 1980s—natural languages

A compiler is a language-translator program that converts the entire program of high-level language into machine language before the computer executes the program. Examples of procedural languages using compilers are COBOL, FORTRAN, Pascal, and C. An interpreter is a language-translator program that converts each procedural language statement into machine language and executes it immediately, statement by statement. BASIC is a procedural language using an interpreter. When a compiler is used, it requires two steps (the source code and the object code) before the program can be executed. The interpreter, on the other hand, requires only one step. The advantage of a compiler language is that, once you have obtained the object code, the program executes faster. The advantage of an interpreter language, on the other hand, is that programs are easier to develop. Some language translators—such as those with C++and Java—can both compile and interpret.

In object-oriented programming, data and the instructions for processing that data are combined into a self-sufficient "object" that can be used in other programs. An object is a self-contained module consisting of preassembled programming code. Send the "message" when an object's data is to be processed. A message is an alert sent to the object when an operation involving that object needs to be performed. The message needs only to identify the operation. How it is actually to be performed is embedded within the processing instructions that are part of the object. These processing instructions within the object are called the methods.

Object-oriented programming (Figure 6-1) has three important concepts, which go under the jaw-breaking names of encapsulation, inheritance, and polymorphism. Encapsulation means an object contains (encapsulates) both data and relevant processing instructions. Once an object has been created, it can be reused in other programs. An object's uses can also be extended through concepts of class and inheritance. All objects that are derived from or related to one another are said to form a class. Each class contains specific instructions (methods) that are unique to that group. Inheritance is the method of passing down traits of an object from classes to subclasses in the hierarchy.

Thus, new objects can be created by inheriting traits from existing classes. Polymorphism means that a message (generalized request) produces different results based on the object that it is sent to.

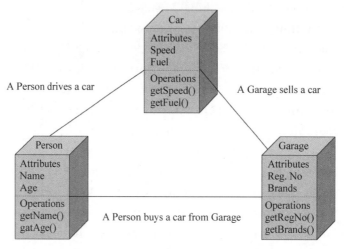

Figure 6-1 Object-oriented Programming

Examples of object-oriented programming languages are C++ & Java. C++— the plus sign **stands for** "more than C"—combines the traditional C programming language with object-oriented capability. C++ was created by Bjarne Stroustrup. With C++, programmers can write standard code in C without the object-oriented features, use object-oriented features, or do a mixture of both. A high-level programming language developed by Sun Microsystems in 1995, Java is used to write **compact** programs that can be downloaded over the Internet and immediately executed on many kinds of computers. Java is similar to C++ but is simplified to eliminate language features that cause common programming errors.

A **markup** language is a kind of coding, or "tags," inserted into text that **embeds** details about the structure and appearance of the text. HTML (HyperText Markup Language) is a markup language that lets people create on-screen documents for the Internet that can easily be linked by words and pictures to other documents. VRML (Virtual Reality Modeling [or Markup] Language) is a type of programming language used to create three-dimensional Web pages including interactive **animation**. XML (eXtensible Markup Language) is a **metalanguage** (a language used to define another language) written in **SGML** that allows one to facilitate the easy interchange of documents on the Internet.

A script is a short list of self-executing commands embedded in a Web page that perform a specific function or routine. JavaScript is a popular object-oriented scripting language that is widely supported in Web browsers. It adds interactive functions to

Unit 6 Implementing the System 系统实现

HTML pages, which are otherwise static. ActiveX is a set of **control**, or reusable components, that enable programs or content of almost any type to be embedded within a Web page. Perl (Practical Extraction and Report Language) is a **general-purpose** programming language developed for text manipulation and now used for Web development, network programming, system administration, **GUI** development, and other tasks.

 Words

instruction[in'strʌkʃn] n. 指令	compiler[kəm'pailə(r)] n. 编译器
statement['steitmənt] n. 语句	interpreter[in'tɜːprətə(r)] n. 解释器, 解释程序
flowchart['fləutʃɑːt] n. 流程图	object['ɔbdʒikt] n. 对象, 目标
code[kəud] v. 编码	self-contained 独立的
letter['letə(r)] n. 字母	jaw-breaking 难发音的, 读起来费劲的
desk-check 手工检查	encapsulation[in͵kæpsju'leiʃn] n. 封装
bug[bʌg] n. 程序缺陷, 臭虫, 漏洞	trait[treit] n. 特性
doubtless['dautləs] adv. 大概, 几乎肯定地	subclass['sʌbklɑːs] n. 子类
surface['sɜːfis] v. 出现, 显露	compact[kəm'pækt, 'kɔmpækt] adj. 简明的, 紧凑的
locate[ləu'keit] v. 找出……的准确位置	markup['mɑːkʌp] n. 标记
typographical[͵taipə'græfikl] adj. 印刷上的, 排字上的	embed[im'bed] v. 嵌入
fine[fain] adv. 够好, 蛮不错	animation[͵æni'meiʃn] n. 动画
end-stage 末期的	metalanguage['metəlæŋgwidʒ] n. 元语言
generation[͵dʒenə'reiʃn] n. 一代, 阶段	control[kən'trəul] n. 控件
assembly[ə'sembli] n. 汇编	general-purpose 通用的

Phrases

in detail 详细地
alpha testing α测试
read through 通读
free of 没有……的
beta testing β测试
go under 归为(某一类事物)中
be derived from 源于, 得自
stand for 表示, 代表

Abbreviations

SGML   Standard for General Markup Language   通用标记语言标准
GUI   Graphical User Interface   图形用户界面

 Exercises

Ⅰ. Read the following statements carefully, and decide whether they are true (T) or false (F) according to the text.

　　____ 1. A program consists of instructions which are composed of statements used in a programming language.
　　____ 2. "Programming" is just "coding" that many people usually think of.
　　____ 3. Sequence, selection, and iteration are the three control structures which are used to form the logic of a program in structured program design.
　　____ 4. Program testing involves two mini steps: desk-checking and debugging.
　　____ 5. A message is a set of processing instructions to perform the operations of the objects involved.

Ⅱ. Choose the best answer to each of the following questions according to the text.

1. Which of the following is right about the steps in the programming process according to this article?
   (A) Problem clarification, program design, program coding, alpha testing, program documentation.
   (B) Problem clarification, algorithm creating, program documentation, program coding, program testing.
   (C) Problem definition, pseudocode, program coding, beta testing, program documentation.
   (D) Problem definition, program design, program coding, program testing, program documentation.

2. Which of the following is right about the program languages according to this article?
   (A) With C++, programmers must use object-oriented features which are different from the standard code in C++.
   (B) All the languages need to be converted by both language-translator programs into machine language to execute.
   (C) Compared to C++, Java is merely simplified to eliminate some complex language features.

(D) Polymorphism allows a message to produce different results based on the object that it is sent to.

3. Which of the following is wrong about the differences between compiler and interpreter?
   (A) A compiler converts the entire program of high-level language into machine language before the computer executes the program.
   (B) An interpreter converts each procedural language statement into machine language and executes it immediately.
   (C) When a compiler requires two steps before the program can be executed, the interpreter requires only one step, so the interpreter enables the program to execute faster.
   (D) An interpreter language enables the program to be developed more easily.

Ⅲ. Fill in the numbered spaces with the words or phrases chosen from the box. Change the forms where necessary.

> lay   understand   develop   base   convert
> possible   come   start   write   build

Design First vs Code First API Development

　　When it ___1___ to using API description formats, two important schools of thoughts have emerged: The "Design First" and the "Code First" approach to API development. The Code First approach is a more traditional approach to ___2___ APIs, with development of code happening after the business requirements are ___3___ out, eventually generating the documentation from the code. The Design First approach advocates for designing the API's contract first before ___4___ any code. This is a relatively new approach, but is fast catching on, especially with the use of API description formats. To ___5___ the two approaches better, let's look at the general process followed during the API lifecycle. Like any product, the concept of the API ___6___ with the business team identifying an opportunity. The opportunity is analyzed and a plan to capitalize on it is created in a text document by strategists, analysts and other business folks. This document is then passed along to the ___7___ team, which is where the plan takes some tangible form. There are two ___8___ from here on to develop the API:

　　1. Design First: The plan is ___9___ to a human and machine readable contract, such as a Swagger document, from which the code is built

　　2. Code First: ___10___ on the business plan, API is directly coded, from which a human or machine readable document, such as a Swagger document can be generated

Ⅳ. **Translate the following passage into Chinese.**
Code Review

Code Review, or Peer Code Review, is the act of consciously and systematically convening with one's fellow programmers to check each other's code for mistakes, and has been repeatedly shown to accelerate and streamline the process of software development like few other practices can. There are peer code review tools and software, but the concept itself is important to understand. Software is written by human beings. Software is therefore often riddled with mistakes. To err is, of course, human, so this is an obvious correlation. But what isn't so obvious is why software developers often rely on manual or automated testing to vet their code to the neglect of that other great gift of human nature: the ability to see and correct mistakes ourselves.

Section B: Good-Enough Software

There's an old(ish) joke about a U.S. company that **places an order** for 100,000 integrated circuits with a Japanese manufacturer. Part of the specification was the defect rate: one chip in 10,000. A few weeks later the order arrived: one large box containing thousands of **ICs**, and a small one containing just ten. Attached to the small box was a label that read: "These are the **faulty** ones."

If only we really had this kind of control over quality. But the real world just won't let us produce much that's truly perfect, particularly not bug-free software. Time, technology, and **temperament** all **conspire against** us.

However, this doesn't have to be frustrating. As Ed Yourdon described in an article in IEEE Software [You95], you can **discipline** yourself to write software that's good enough—good enough for your users, for future maintainers, for your own **peace of mind**. You'll find that you are more productive and your users are happier. And you may well find that your programs are actually better for their shorter **incubation**.

Before we go any further, we need to qualify what we're about to say. The phrase "good enough" does not imply **sloppy** or poorly produced code. All systems must meet their users' requirements to be successful. We are simply advocating that users be given an opportunity to participate in the process of deciding when what you've produced is good enough (Figure 6-2).

Figure 6-2 Good Enough

Normally you're writing software for other people. Often you'll remember to get requirements from them. But how often do you ask them how good they want their software to be? Sometimes there'll be no choice. If you're working on **pacemakers**, the space shuttle, or a low level library that will be widely **disseminated**, the requirements will be more **stringent** and your options more limited. However, if you're working on a brand new product, you'll have different constraints. The marketing people will have promises to keep, the eventual end users may have made plans based on a delivery schedule, and your company will certainly have cash-flow constraints. It would be unprofessional to ignore these users' requirements simply to add new features to the program, or to **polish up** the code just one more time. We're not advocating panic; it is equally unprofessional to promise impossible time scales and to **cut** basic engineering **corners** to meet a deadline.

The scope and quality of the system you produce should be specified as part of that system's requirements.

Often you'll be in situations where trade-offs are involved. Surprisingly, many users would rather use software with some rough **edges** today than wait a year for the multimedia version. Many IT departments with tight budgets would agree. Great software today is often preferable to perfect software tomorrow. If you give your users something to play with early, their feedback will often lead you to a better eventual solution.

In some ways, programming is like painting. You start with a blank canvas and certain basic raw materials. You use a combination of science, art, and craft to determine what to do with them. **You sketch out** an overall shape, paint the underlying environment, then fill in the details. You constantly **step back** with a critical eye to view what you've done. **Every now and then** you'll throw a canvas away and start again.

But artists will tell you that all the hard work is ruined if you don't know when to stop. If you add layer upon layer, detail over detail, the painting becomes lost in the paint.

Don't **spoil** a perfectly good program by over-**embellishment** and over-refinement. Move on, and let your code stand in its own right for a while. It may not be perfect. Don't worry; it could never be perfect.

Words

faulty ['fɔːlti] adj. 有错误的；有缺陷的
temperament ['temprəmənt] n. 气质, 性情, 性格
discipline ['disəpln] v. 自我控制, 严格要求（自己）
incubation [ˌiŋkjuˈbeʃn] n. 培育
sloppy ['slɒpi] adj. 草率的, 粗心的
pacemaker ['peismeikə(r)] n. 起搏器

disseminate [diˈsemineit] v. 传播（信息、知识等）
stringent ['strindʒənt] adj. 严格的, 严厉的
edge [edʒ] n. 优势
spoil [spɔil] v. 破坏, 糟蹋
embellishment [imˈbeliʃmənt] n. 修饰, 润色

Phrases

```
place an order    订购,下单,发出订单
if only           要是……多好,只要
conspire against  密谋反对
peace of mind     内心的宁静,心如止水
polish up         改善,润色,使完美
cut corners       抄近路,以简便方法做事
sketch out        草拟,概略地叙述
step back         后退,退后
every now and then 不时地,常常
```

 Abbreviations

IC   Integrated Circuit   集成电路,芯片

 Exercises

Ⅰ. Read the following statements carefully, and decide whether they are true (T) or false (F) according to the text.

　　____ 1. You may spoil a perfectly good program by over-embellishment and over-refinement.

　　____ 2. The phrase "good enough" imply sloppy or poorly produced code.

　　____ 3. It would be professional to ignore these users' requirements simply to add new features to the program, or to polish up the code just one more time.

　　____ 4. Usually you're writing software for other people.

　　____ 5. It is professional to promise impossible time scales and to cut basic engineering corners to meet a deadline.

Ⅱ. Choose the best answer to each of the following questions according to the text.

1. Which of the following is described by Ed Yourdon in an article in IEEE Software [You95] regarding the good enough software?

(A) Good enough for your users

(B) Good enough for future maintainers

(C) Good enough for your own peace of mind

(D) All of the above

Unit 6 Implementing the System 系统实现

2. Which of the following is right about good enough software you will or may find according to this article?

   (A) You are more productive.

   (B) Your users are happier.

   (C) Your programs are actually better for their shorter incubation.

   (D) All of the above

3. Which of the following is mentioned in this article?

   (A) Grady Booch

   (B) James Rumbaugh

   (C) Ed Yourdon

   (D) Ivar Jacobson

Ⅲ. Fill in the numbered spaces with the words or phrases chosen from the box. Change the forms where necessary.

> from without play maintain change
> apply clean appear throughout cover

Refactoring

Refactoring is "the process of ___1___ a software system in such a way that it does not alter the external behavior of the code yet improves its internal structure," according to Martin Fowler, the "father" of refactoring. The concept of refactoring ___2___ practically any revision or cleaning up of source code, but Fowler consolidated many best practices ___3___ across the software development industry into a specific list of "refactorings" and described methods to implement them in his book, Refactoring: Improving the Design of Existing Code.

One approach to refactoring is to improve the structure of source code at one point and then extend the same changes systematically to all applicable references ___4___ the program. The result is to make the code more efficient, scalable, maintainable or reusable, ___5___ actually changing any functions of the program itself. In his book, Fowler describes a methodology for ___6___ up code while minimizing the chance of introducing bugs.

While refactoring can be ___7___ to any programming language, the majority of refactoring current tools were originally developed for the Java language. In 2001, automated refactoring tools began to ___8___ in earnest, including the IntelliJ IDEA Java IDE (Integrated Development Environment), the X-ref plug-in tool for the Emacs editor and the Instantiations jFactor stand-alone refactoring tool.

Many basic editing environments support simple refactorings like renaming a function or variable across an entire code base. Eric Raymond, a leading philosopher about program development, ___9___ that the concept of refactoring is consistent with the idea of get-something-working-now-and-perfect-it-later approach long familiar to Unix and open source programmers and hackers. The idea is also embodied in the approach known as extreme programming.

Today, refactoring ___10___ an important role in application modernization and moving legacy apps from a monolithic structure to microservices.

Ⅳ. **Translate the following passage into Chinese.**
Software Complexity

Grady Booch (one of the co-authors of the Unified Modeling Language), in his book "Object Oriented Design", describes in the very first chapter why the world is complex by nature (with all its events and its properties) and consequently the software has to be complex since it is a representation of reality. However, a point that Booch neglects to specify is that such complexity can be very well abstracted so that end users don't have to deal with it. In other words, Booch makes reference to the human capability of abstraction just to explain it as a brain process that is modeled in the Object Oriented programming paradigm; but does not relate it to his first chapter of complexity.

In short: Despite reality is complex by nature, and software is a model of the reality, software can be simplified by building a foundation framework from which new models can be created; that hides most complexity and allows users of the higher layers to focus only on the semantics of the problem they intend to solve.

Part 3

Simulated Writing: Progress Report

Functions and Contents of Progress Reports

Progress reports are common in engineering. Once you have written a successful proposal and have secured the resources to do a project, you are expected to update the client on the progress of that project. This updating is usually handled by progress reports. They might be one-page memos or long, letters, short reports, formal reports, or presentations. Such a report is aimed at whoever assigned the project. Your goals are to give a fair assessment of what you have accomplished on the project and of what you expect to do in the future, and to enable the manager or sponsor of a project to make informed decisions about the future of the project. Yet, any project of size or significance is bound to encounter snags: additional requirements, miscommunications, problems, delays, or unexpected expenses. A progress report must account for those snags. You should discuss any problems you are having and what further help you require.

What information expected in a progress report depends on the situation, but most progress reports have the following similarities in content:

(1) Background on the project itself.

(2) Discussion of achievements since last reporting.

(3) Discussion of problems that have arisen.

(4) Discussion of work that lies ahead.

(5) Assessment of whether you will meet the objectives in the proposed schedule and budget.

Organizational Patterns for Progress Reports

The structure of the progress report is determined by the original proposal for the project: make use of original milestones or the timeline. With this in mind, a more comprehensive list of components will give you a clearer structure as follows.

(1) Introduction

(2) Project Description

(3) Work Completed

(4) Problems Arisen

(5) Changes in Requirements

(6) Work Scheduled

(7) Overall Assessment of the Project

1. Introduction

As always, first indicate the purpose of the report and its intended audience. Clearly define the time period covered in the report. Then, explain what your project is, what its objectives are, and what the status of the project was at the time of the last reporting.

2. Project Description

In very short reports, the introduction might contain this section, but if it is under its own heading, readers who are familiar with the project can skip it. Someone unfamiliar with the project, however, needs summarized details such as purpose and scope of the project, start and completion dates, and names of parties involved.

3. Work Completed

This section follows the progress of the tasks presented in the proposal's schedule. This is the substance of the report. This section would be a project-tasks approach, a time-periods approach or a combined approach.

- Project-tasks approach: Focus on the tasks. Defined milestones can logically organize your discussion into this kind of structure. Also if you are working on a number of semi-independent tasks at the same time, this approach will work well.
- Time-periods approach: Focus on time. If a timeline (or deadline) is more important than milestones, then use this approach. Also, use it for projects with a simple linear structure.
- Combined approach: The two above approaches could be combined if, for example, under previous work, you break down what you have done by individual tasks. Or, under the tasks, you focus on what part is complete, what part is in progress, and what part is yet to come.

Your project will determine which of these three you use. If the problems encountered or changes required are time-related, then use the time-periods approach to your advantage; likewise, if the problems or changes relate to specific tasks then use the project-tasks approach. Another item that may be included here is a summary of financial data. This last item could be contained in a table or appendix, or an independent section.

4. Problems Arisen

Progress reports are not necessarily for the benefit of only the client. As noted in the opening, snags are expected. Don't hide from them; explain what they are and how they might affect key areas of the job (such as timing, price or quality). Often, you the engineer or scientists benefit from the reporting because you can share or warn your client about problems that have arisen. If the problem occurred in the past, you can explain how you overcame it. This is least serious; in fact, you look good. If the problem is in front of you (now or in the future), explain how you hope to overcome it, if you can.

5. Changes in Requirements

Here, you record the changes to the project: milestones added, new requirements, or schedule changes (good or bad). Even if these changes have not affected the ultimate goal of the project, you need to tell the sponsor how problems have been accommodated.

Note: If changes are a direct result of problems encountered, sections 4 and 5 may be combined. This would lead to a modified organization: first problem and the change it required, then the next problem and change, and so on.

6. Work Scheduled

In this section, you discuss your plan for meeting the objectives of the project. In many ways, this section of a progress report is written in the same manner as the "Plan of Action" section of the proposal, except that now you have a better perspective for the schedule and cost than you did earlier. You can use one of those three approaches mentioned above (a project-tasks approach, a time-periods approach or a combined approach) according to your project in this section like in the section "Work Completed".

7. Overall Assessment of the Project

Since a progress report is not about a finished work, the conclusion needs only to give your professional opinion of how the project is going. Being unrealistically optimistic is as inappropriate as being unduly negative. Beware of promising early completion: a single setback can gobble up much time. Likewise, don't overreact if you are behind schedule. You may also gain time along the way. Far more significant for the engineer is to explain anything that may change the expected quality of the final product. Keeping in mind your purpose can help you focus here: Your goal is to enable the manager or sponsor to make informed decisions.

P. S. This guide can be tailored according to the different progress reports.

A Sample Progress Report

TO: Steven Zhang
Departments of Computer Science, Beihang University
FROM: Mark Hu
SUBJECT: Progress on Scheme Debugging Report
DATE: August 3, 2020

This memo describes the progress I have made to date on my independent-study project to write a report on debugging in Scheme. In this memo, I review the nature of the project and describe work I have completed, work I am currently engaged in, and work I plan to complete by the end of the project.

As I described in my memo of July 4, this project will result in a technical report whose purpose is to provide readers with practical information on developing and debugging programs in Scheme, supplementary to the material in your textbook, An Introduction to Scheme and its Implementation.

Project Description

The report is aimed at students in computer science (undergraduate and graduate) who have previous programming experience, but are new to Scheme. The information in this report is needed because readers who have developed programs using compilers for other languages may be unfamiliar with the approaches available with an interactive interpreter and debugger.

Project Scope

In my earlier memo, I proposed to cover the following high-level topics:

(1) Loading the debugging module into the interpreter

(2) Establishing break levels

(3) Applying back-trace

(4) Managing dependencies

(5) Saving and loading a customized heap image of the Scheme system

(6) Debugging local definitions

(7) Debugging native-code procedure calls

(8) Debugging when using functional programming style

(9) Program design and implementation strategies

(10) Using stubbed procedures

(11) Differences between RScheme and other Scheme systems

In my current outline, these are divided into three major parts, with an addendum for topic 11. The three parts are: (A) basic debugging procedures—topics 1-3, (B) advanced debugging procedures—topics 4-7, and (C) general program development strategies—topics 8-10.

Work Completed

I have completed first drafts of the sections in part A on loading the debugging module, break levels, and apply-back-trace. I intend to make note of additional material for these sections while working on the later sections, if further background information is needed.

Present Work

I am currently working on the sections in part B. Since these sections are highly interrelated, I am working on them roughly in parallel. I am also currently researching information on other Scheme systems for section 11; I have located information on Gambit and DrScheme. I expect the current work to be completed by the end of this week, August 9th.

Future Work

Next, I will draft the sections in part C and the addendum on other Scheme systems. Finally, I will fully revise the entire draft, integrating further material where deficiencies have become evident during work on other sections. The final report will be ready for your review on August 20th.

Conclusion

Thus far, the project is proceeding well. I have not run into any major problems, nor do I anticipate any in the remaining work.

 Exercises

1. What is this progress report mainly talking about?
2. Based on the project you are doing, write a similar progress report.

Unit 7

Testing the System

系统测试

Unit 7 Testing the System 系统测试

Part 1

Listening & Speaking

Unit 7

Dialogue: Software Testing

Kevin: It seems that we haven't spent enough time on talking about testing.

Sharon: True, but we have been thinking about it. In fact, more than thinking. Our testing strategy planned in the requirements phase, **takes an** incremental **view of** testing, beginning with the testing of individual program units, moving to testing designed to facilitate the integration of the units, and **culminating** with testing that **exercises** the constructed system.

Kevin: Yes. We have completed the design of the **unit testings** for each component of our system before coding begins, which is a preferred approach by agile programming. Even though we are not using eXtreme Programming, we decided that it would be a good idea—a review of design information provides guidance for establishing test cases that are likely to **uncover** errors in each of the categories discussed earlier. Now, we can conduct our unit testing using these test cases.

Jason: During the **integration testing** phase, we can use some classical integration test strategies, such as top-down or bottom-up. Besides those used for traditional, hierarchically structured software, we can also use architecture-driven, which implies integrating the software components or subsystems based on identified

Jason: functional **threads** according to the requirements specification.[1]

Kevin: **Well done**. During the integration testing, we can develop a validation strategy at the same time. In this step, we should define testing required for validation and its criteria according to the software requirements specification, and information contained in its Validation Criteria section forms the basis for a **validation testing** approach.

Sharon: Maybe we have to perform some high-level testing such as recovery testing, security testing, **stress testing** and **performance testing**. Finally, we will coordinate **acceptance testing** with the customer.

[1] Replace with:
1. We can also use **black-box-box testing** when testing the input/output behavior of system.
2. We can also use **regression testing**, …

Exercises

Work in a group, and make up a similar conversation by replacing the statements with other expressions on the right side.

Words

culminate [ˈkʌlmineit] v. 告终，达到顶点
exercise [ˈeksəsaiz] v. 执行，使用
uncover [ʌnˈkʌvə(r)] v. 发现,揭开,揭露
thread [θred] n. 各组成部分

Phrases

takes a view of 观察,检查
unit testing 单元测试
integration testing 集成测试
black-box testing 黑盒测试
regression testing 回归测试
well done 做得好
validation testing 确认测试
stress testing 压力测试

performance testing 性能测试
acceptance testing 验收测试

Listening Comprehension: Software Testing

Listen to the article and answer the following 3 questions based on it. After you hear a question, there will be a break of 15 seconds. During the break, you will decide which one is the best answer among the four choices marked (A), (B), (C) and (D).

Questions

1. Which kind of approaches is also called "functional testing"?
   (A) White-box testing
   (B) Black-box testing
   (C) Grey-box testing
   (D) State-based testing

2. Which kind of approaches does the mutation testing belong to?
   (A) White-box testing
   (B) Black-box testing
   (C) Grey-box testing
   (D) State-based testing

3. Which testing strategy is generally wrong according to this article?
   (A) Using white-box testing for unit testing
   (B) Using black-box testing for integration testing.
   (C) Using white-box testing for system testing
   (D) Using black-box testing for acceptance testing

Words

verification[ˌverifiˈkeiʃn] n. 验证,确认,验收
introduce[ˌintrəˈdjuːs] v. 引入
deduce[diˈdjuːs] v. 推论,演绎出
mutant[ˈmjuːtənt] n. 变体

Phrases

equivalence class partitioning 等价类划分法
boundary value analysis 边界值分析法
cause-effect graphing 因果图法
state-based testing 基于状态的测试

grey-box testing　灰盒测试
white-box testing　白盒测试
mutation testing　变异测试,突变测试

Dictation：Smoke Test

This article will be played three times. Listen carefully, and fill in the numbered spaces with the appropriate words you have heard.

In electronics, the first time a circuit is attached to power, which will sometimes produce actual smoke if a design or wiring mistake has been made, is called "___1___ testing". It refers to the first test made after repairs or first ___2___ to provide some assurance that the system under test will not **catastrophically** fail.

In software, the term smoke testing is **metaphorical** which describes the process of ___3___ code changes before the changes are ___4___ into the product's source tree. After code reviews, smoke testing is the most cost effective method for identifying and fixing ___5___ in software. It is designed to ___6___ that changes in the code function as ___7___ and do not **destabilize** an entire **build**.

McConnell, the author of Code Complete, ___8___ that, "Whatever ___9___ strategy you select, a good approach to integrating the software is the 'daily ___10___ and smoke test'." Every file is compiled, linked, and combined into an ___11___ program as a "build" every day, and the program is then **put through** a "smoke test," a relatively simple check to exercise the entire system **from end to end**. It does not have to be exhaustive, but it should be capable of ___12___ major problems that will keep the build from properly performing its function.

This simple process provides a number of significant ___13___ when it is applied on complex, time-___14___ software engineering projects. Firstly, it minimizes the integration ___15___ by ___16___ on all the code daily and ___17___ quality problems early. It also simplifies defect diagnosis and ___18___ because the software that has just been added to the build(s) is a probable cause of a newly discovered error. Besides, it supports easier progress ___19___ and improves **morale** when developers seeing ___20___ more of the product works every day with daily builds.

▸ Words

catastrophically[ˌkætəˈstrɒfikli] *adv.* 灾难性地

metaphorical[ˌmetəˈfɒrikl] *adj.* 隐喻性的,比喻性的

Unit 7　Testing the System　系统测试

destabilize[ˌdiːˈsteibəlaiz] v. 使不安定，使不稳定

build[bild] n. 编好的程序，构造

morale[məˈrɑːl] n. 士气，斗志

 Phrases

smoke test　冒烟测试
put through　完成
from end to end　从头到尾地

Part 2

Reading & Translating

Section A: Software Testing

Testing is an activity performed for evaluating product quality, and for improving it, by identifying defects and problems. The view of software testing has evolved towards a more constructive one (Figure 7-1). Testing is no longer seen as an activity which starts only after the coding phase is complete, with the limited purpose of detecting failures. Software testing is now seen as an activity which should encompass the whole development and maintenance process and is itself an important part of the actual product construction. Indeed, planning for testing should start with the early stages of the requirement process, and test plans and procedures must be systematically and continuously developed, and possibly refined, as development **proceeds**. These test planning and designing activities themselves constitute useful input for designers in **highlighting** potential weaknesses (like design **oversights** or contradictions, and **omissions** or ambiguities in the documentation).

Software testing is usually performed at different levels along the development and maintenance processes. **That is to say**, the target of the test can vary: a single module, a group of such modules (related by purpose, use, behavior, or structure), or a whole system. Three big test stages can be conceptually distinguished, namely Unit, Integration, and System. No process model is implied, nor are any of those three stages assumed to have greater importance than the other two.

Unit testing

Unit testing verifies the **functioning in isolation of** software pieces which are separately testable. Depending on the context, these could be the individual subprograms or a larger component made of tightly related units. Typically, unit testing occurs **with**

Figure 7-1　Software Testing

access to the code being tested and with the support of debugging tools, and might involve the programmers who wrote the code.

Integration testing

Integration testing is the process of verifying the interaction between software components. Classical integration testing strategies, such as top-down or bottom-up, are used with traditional, hierarchically structured software. Modern systematic integration strategies are rather architecture-driven, which implies integrating the software components or subsystems based on identified functional threads.

System testing

System testing is concerned with the behavior of a whole system and is usually considered appropriate for comparing the system to the non-functional system requirements, such as security, speed, accuracy, and reliability. External interfaces to other applications, utilities, hardware devices, or the operating environment are also evaluated at this level.

Other important objectives for testing include (but are not limited to) usability evaluation, and acceptance, for which different approaches would be taken. Note that the test objective varies with the test target; in general, different purposes being addressed at a different level of testing.

Acceptance/qualification testing

Acceptance testing checks the system behavior against the customer's requirements, however these may have been expressed; the customers undertake, or specify, typical tasks to check that their requirements have been met or that the organization has identified these for the target market for the software. This testing activity may or may not involve the developers of the system.

Unit 7 Testing the System 系统测试

Alpha and beta testing

Before the software is released, it is sometimes given to a small, representative set of potential users for trial use, either in-house (alpha testing) or external (beta testing). These users report problems with the product. Alpha and beta use is often uncontrolled, and is not always referred to in a test plan.

Many techniques have been developed to perform testing, which attempt to "break" the program, by running one or more tests drown from identified classes of executions deemed equivalent.[1] The leading principle underlying such techniques is to be systematic as possible in identifying a representative set of program behaviors; for instance, considering subclasses of the input domain, scenarios, states, and dataflow. It is difficult to find a homogeneous basis for classifying all techniques, and the one used here must be seen as a compromise. The classification is based on how tests are generated from the software engineer's institution and experience, the specification, the code structure, the (real or artificial) faults to be discovered, the field usage, or finally, the nature of the application. Sometimes these techniques are classified as white-box, also called glass-box, if the tests rely on information about how the software has been designed or coded, or as black-box if the test cases rely on the input/output behavior. One last category deals with combined use of two or more techniques. Obviously, these techniques are not used equally often by all practitioners.

 Words

proceed[prəˈsiːd] v. 继续进行	deem[diːm] v. 认为,视作
highlight[ˈhailait] v. 强调,使突出	homogeneous[ˌhoməˈdʒiːniəs] n. 相同特征的,同性质的
oversight[ˈəuvəsait] n. 疏忽,漏失	
omission[əˈmiʃn] n. 疏忽,遗漏	compromise [ˈkɔmprəmaiz] n. 妥协,折中
functioning[ˈfʌŋkʃəniŋ] n. 功能	
undertake[ˌʌndəˈteik] v. 着手做,开始进行	institution[ˌinstiˈtjuːʃn] n. 习俗,制度
	practitioner[prækˈtiʃənə(r)] n. 从业者,实践者
release[rˈliːs] v. 发布,公开发行	
in-house 内部的	

 Phrases

that is to say	即,换句话说
in isolation of	脱离
with access to	访问

be concerned with　涉及,关于,关心,关注
compare...to　把……比作
check against　对照
trial use　试用,试行
be referred to　被提及,涉及

Notes

[1]　**Original**：Many techniques have been developed to perform testing, which attempt to "break" the program, by running one or more tests drown from identified classes of executions deemed equivalent.

　　Translation：很多技术已被开发用于执行测试,通过运行一个或多个从已确定的被认为是等价执行类中获得的测试,试图"破坏"程序。

Exercises

Ⅰ. Read the following statements carefully, and decide whether they are true（T）or false（F）according to the text.

　　____ 1. Test plan should initiate in the early stage of a software project.
　　____ 2. Developers of the system only need to participate in the first three testing activities.
　　____ 3. Top-down or bottom-up are two classical integration testing strategies.
　　____ 4. Test plan is helpful for preventing design defects.
　　____ 5. Glass-box techniques are used to test the inner structures of software.

Ⅱ. Choose the best answer to each of the following questions according to the text.

　　1. Which of the following is wrong about the software testing?
　　　（A）Software testing should cover maintenance process besides development process.
　　　（B）Test plan should be completed before coding.
　　　（C）Test planning and designing activities can forecast potential weaknesses.
　　　（D）The target of the testing can vary at different levels and phases.

　　2. Which of the following is right about the testing activities?
　　　（A）Unit testing begins when all the units have been completed.
　　　（B）System testing checks whether the system meets the customer's non-functional requirements.
　　　（C）Acceptance testing checks whether the system meets the customer's

Unit 7 Testing the System 系统测试

functional requirements.

（D）Before the software is released，it is sometimes given to a small，random set of potential users for trial use.

3. Which of the following is right about the testing techniques?

（A）The testing techniques perform testing by running one or more tests that are equivalent to identified classes of executions.

（B）The leading testing principle is to identify a representative set of program behaviors.

（C）All the testing techniques are classified based on a homogeneous standard.

（D）Black-box techniques do not care how the software has been designed or coded.

Ⅲ. Fill in the numbered spaces with the words or phrases chosen from the box. Change the forms where necessary.

> conduct   carry   call   interpret   meet
> common   main   base   involve   discover

Alpha Test

An alpha test is a preliminary software field test ___1___ out by a team of users in order to find bugs that were not found previously through other tests. The ___2___ purpose of alpha testing is to refine the software product by finding (and fixing) the bugs that were not ___3___ through previous tests.

The team that ___4___ the alpha test is often an independent test team, perhaps made up of potential users/customers. Alpha testing ___5___ simulating a real user environment by carrying out tasks and operations that the actual users might perform. Once software passes the alpha test, it is considered for the next phase of testing ___6___ beta testing.

The meaning of alpha also can differ ___7___ on whether the project is custom software done for a client. In this case, alpha testing implies an initial meeting between software vendor and client to ensure that the client's requirements are properly ___8___ by the developer in terms of the performance, functionality and durability of the software program.

Compare this versus the context of a Web application, where alpha testing can be ___9___ as an online application that isn't completely ready for usage, but that has been opened up to get some initial feedback. ___10___ an alpha might allow power users to get their first look at the system via a private invitation.

Ⅳ. Translate the following passage into Chinese.

Conventional Testing vs. Object Oriented Testing

Conventional testing is the traditional approach to testing mostly done when water fall life cycle is used for development, while object oriented testing is used when object oriented analysis and design is used for developing enterprise software. Conventional testing focuses more on decomposition and functional approaches as opposed to object oriented testing, which uses composition. The three levels of testing (system, integration, unit) used in conventional testing is not clearly defined when it comes to object oriented testing. The main reason for this is that OO development uses incremental approach, while traditional development follows a sequential approach. In terms of unit testing, object oriented testing looks at much smaller units compared to conventional testing.

Section B: Testers and Programmers Working Together

Let's look at an example of how a tester and programmer might work on a user story or feature. Patty Programmer and Tammy Tester are working on a user story to calculate the shipping cost of an item, based on weight and destination postal code. Tammy writes a simple test case in a **tabular** format that is supported by their Fit-based test tool (Table 7-1):

Table 7-1 A Simple Test Case

Weight	Destination Postal Code	Cost
5 kg	80104	$ 7.25

Meanwhile, Patty writes the code to send the inputs to the shipping cost API and to get the calculated cost. She shows Tammy her unit tests, which all pass. Tammy thinks Patty's tests look ok, and they agree Patty will check in the code.

Next, Patty checks in a **fixture** to automate Tammy's tests. Patty **calls** Tammy **over** to show her that the first simple test is working. Tammy **writes** up more test cases, trying different weights and destinations within the U.S. Those all work fine. Then she tries a Canadian postal code, and the test **dies** with an exception. She shows this to Patty, who realizes that the shipping cost calculator API **defaults to** U.S. postal codes, and requires a country code for postal codes in Canada and Mexico. She hadn't written any unit tests for any other countries yet.

Tammy and Patty **pair** to revise the inputs to the unit tests. Then Patty pairs with Carl Coder to change the code that calls the API. Now the test looks like Table 7-2.

Table 7-2　A Test Case

Weight	Destination Postal Code	Country Code	Cost
5 kg	80104	US	$ 7.25
5 kg	T2J 2M7	CA	$ 9.40

　　This **back-and-forth** testing and coding process could take all kinds of forms. Patty might write these "story tests" herself, in addition to her unit tests. Or, she and Tammy may decide that they can cover all of Tammy's acceptance tests with unit-level tests. Patty might be in a remote office, using an online collaboration tool to pair with Tammy. Either or both might pair with other team members. They might need help from their database expert to set up the test database. **The point is** that testing and coding are part of one process, in which all team members participate.

　　Tammy can keep identifying new test cases until she feels all the risky areas have been covered. She might test with the **heaviest** possible item, and the most expensive destination. She might test having a large quantity of one item, or many items to the same destination. Some **edge cases** might be so unlikely she doesn't bother with them. She may not keep all the automated tests in the regression test suite. Some tests might be better done manually, after a UI is available.

　　Patty has written unit tests with Hawaii as the shipping destination, but Tammy thinks only continental destinations are acceptable. They both go to talk to the product owner about it. This is the Power of Three. When questions arise, having three different viewpoints is an effective way to make sure you get the right solution, and you don't have to **rehash** the issue later. This helps prevent requirement changes from flying in **under the radar** and causing unpleasant surprises later.

　　It's vital that everyone on the development team understands the business, so don't fall into the habit of only having a tester, an analyst or a programmer communicate with the business experts.

　　The story about how Tammy and Patty work together shows how closely programmers and testers collaborate. As coding and testing proceed, there are many opportunities to transfer skills. Programmers learn new ways of testing. Testers learn more about code design and how the right tests can improve it.

　　Patty has completed the UI for selecting shipping options and displaying the cost, but hasn't checked it in yet. She calls Tammy over to her workstation and demonstrates how the end user would enter the destination postal code, select the shipping option, and see the cost right away. Tammy **tries** this **out**, changing the postal code to see the new cost appear. She notices the **text box** for the postal code allows the user to enter more characters than should be allowed for a valid code, and Patty changes the html accordingly. Once the UI looks good, Patty checks in the code, and Tammy continues with her exploratory testing.

Tammy is especially concerned with changing the postal code and having the new cost display, as they identified this as a risky area. She finds that if she displays the shipping cost, goes on to the next page of the UI, then comes back to change the postal code, the new estimated cost doesn't display. She asks Patty to come observe this behavior. Patty realizes there is a problem with values being **cached**, and goes back to her workstation to fix it.

Showing someone a problem real-time is much more effective than filing a bug in a defect tracking system and waiting for someone to have time to look at it later. If the team is distributed and people are in different time zones, it's harder to do **work through** issues together. The team will have to make adjustments to get this kind of value. One of your teammates is in a time zone 12.5 hours ahead, but works late into his nighttime to overlap with your morning. You work through test results and examples when you're both online.

Show the customers, too. As soon as you have a prototype, some basic navigation, some small testable piece of code, show it to the customer and get their feedback. Feedback, from our customers, from our automated tests, from each other, is our most powerful tool in staying on track and delivering the right business value.

When we divide our work into small, manageable chunks, plan and conduct testing and coding as part of a single development process, and focus on finishing one chunk of valuable functionality at a time, testing doesn't **get squeezed to** the end, **put off** to a future iteration, or ignored altogether.

Get your team together today and talk about how you can all—testers, programmers and everyone else involved with delivering the software—work together to integrate coding and testing. Instead of investing in a big requirements document, capture requirements and examples of desired application behavior in executable tests, and write the code that will make those pass. Meet with your business stakeholders to understand their priorities and explain how much work you can realistically take on each iteration. Stop treating coding and testing as separate activities (Figure 7-2).

Figure 7-2　Stop Treating Coding and Testing as Separate Activities

It won't happen overnight, but gradually your team will get better and better at really finishing each software feature—including all the testing. Your customers will be delighted to get stable, robust software that meets their needs. Your team will benefit from better-designed code that's easier to maintain and contains far fewer bugs. **Best of all**, testers and programmers **alike** will enjoy their work much more!

 Words

tabular['tæbjələ(r)] *adj.* 列成表格的
fixture['fɪkstʃə(r)] *n.* 固定装置
die[daɪ] *v.* 停止运转
pair[peə(r)] *v.* 使成对,配对
back-and-forth 反复地,来回地
heaviest['hevɪst] *adj.* (在数量、程度等方面)超出一般的

rehash['riːhæʃ] *v.* 事后反复回想(或讨论)
cache[kæʃ] *v.* 隐藏,缓存
alike[ə'laɪk] *adv.* 同样地

 Phrases

call over   把……叫过来
write up   详细写出
default to   默认为
the point is 问题在于
edge case   边界用例,极端例子
under the radar   避开别人关注的行为,低调处理
try out   试验,尝试
text box   正文框
work through   解决,完成,干完
get squeezed to   陷入,挤到
put off   推迟
best of all   首先,最好的是,最重要的是

Exercises

Ⅰ. Read the following statements carefully, and decide whether they are true (T) or false (F) according to the text.

　　____ 1. Tammy is a Programmer and Patty is a Tester.
　　____ 2. Patty tries a Canadian postal code, and the test dies with an exception.
　　____ 3. Carl is a Coder.
　　____ 4. Patty and Tammy may decide that they can cover all of Tammy's acceptance

tests with unit-level tests.

_____ 5. As coding and testing proceed, there are not many opportunities to transfer skills.

Ⅱ. **Choose the best answer to each of the following questions according to the text.**

1. Which of the following is wrong about Tammy?
   (A) Tammy writes a simple test case in a tabular format that is supported by their Fit-based test tool.
   (B) Tammy is a Tester.
   (C) Tammy is a Programmer.
   (D) All of the above.

2. Which of the following is right about Patty?
   (A) Patty is a Tester.
   (B) Patty is a Programmer.
   (C) Patty writes the code to send the inputs to the shipping cost API and to get the calculated cost.
   (D) None of the above.

3. Which of the following is wrong about programmers and testers?
   (A) Programmers learn new ways of testing.
   (B) Testers learn more about code design and how the right tests can improve it.
   (C) The story about how Tammy and Patty work together shows how closely programmers and testers collaborate.
   (D) None of the above.

Ⅲ. **Fill in the numbered spaces with the words or phrases chosen from the box. Change the forms where necessary.**

> release   make   available   use   which
> refer   include   subject   consider   original

Beta Test

In software development, a beta test is the second phase of software testing in ___1___ a sampling of the intended audience tries the product out.

Beta is the second letter of the Greek alphabet. ___2___, the term alpha test meant the first phase of testing in a software development process. The first phase ___3___ unit testing, component testing, and system testing. Beta testing can be

___4___ "pre-release testing."

Beta testing is also sometimes ___5___ to as User Acceptance Testing (UAT) or end user testing. In this phase of software development, applications are ___6___ to real world testing by the intended audience for the software. The experiences of the early users are forwarded back to the developers who make final changes before ___7___ the software commercially.

For in-house testing, volunteers or paid test subjects ___8___ the software. For widely-distributed software, developers may make the test version ___9___ for downloading and free trial over the Web. Another purpose of ___10___ software widely available in this way is to provide a preview and possibly create some buzz for the final product.

Ⅳ. **Translate the following passage into Chinese.**

Regression Testing

Whenever developers change or modify their software, even a small tweak can have unexpected consequences. Regression testing is testing existing software applications to make sure that a change or addition hasn't broken any existing functionality. Its purpose is to catch bugs that may have been accidentally introduced into a new build or release candidate, and to ensure that previously eradicated bugs continue to stay dead. By re-running testing scenarios that were originally scripted when known problems were first fixed, you can make sure that any new changes to an application haven't resulted in a regression, or caused components that formerly worked to fail. Such tests can be performed manually on small projects, but in most cases repeating a suite of tests each time an update is made is too time-consuming and complicated to consider, so an automated testing tool is typically required.

Part 3

Simulated Writing: Software Test Specification

The test plan describes an overall breakdown of testing into individual tests that address specific items. For each such individual test, we write a test specification and evaluation. The specification begins by listing the requirements whose satisfaction will be demonstrated by the test. Referring to the requirements documents, this section explains the test's purpose.

One way to view the correspondence between requirements and tests is to use a table or chart. The system functions involved in the test are enumerated in the table. The performance tests can be described in a similar way. Instead of listing functional requirements, the chart lists requirements related to speed of access, database security, and so on.

Often, an individual test is really a collection of smaller tests, the sum of which illustrates requirements satisfaction. In this case, the test specification shows the relationship between the smaller tests and the requirements.

Each test is guided by a test philosophy and adopts a set of methods. However, the philosophy and methods may be constrained by other requirements and by the realities of the test situation. The specification makes these test conditions clear. Among the conditions may be some of the following:

- Is the system using actual input from users or devices, or are special cases generated by a program or surrogate device?
- What are the test coverage criteria?
- How will data be recorded?
- Is there any timing, interface, equipment, personnel, database, or other limitations on testing?
- If the test is a series of smaller tests, in what order are the tests to be performed?

If test data are to be processed before being evaluated, the test discusses the processing. For instance, when a system produces large amounts of data, data reduction techniques are sometimes used on the output so that the result is more suitable for evaluation.

Accompanying each test is a way to tell when the test is complete. Thus, the specification is followed by a discussion of how we know when the test is over and the relevant requirements have been satisfied. For example, the plan explains what range of output results will meet the requirement.

The evaluation method follows the completion criteria. For example, data produced during testing may be collected and collated manually and then inspected by the test team. Alternatively, the team could use an automated tool to evaluate some of the data and then inspect summary reports or do an item-by-item comparison with expected output.

Software Test Specification Template

Unit 7 Testing the System 系统测试

1. Introduction

This section provides an overview of the entire test document. This document describes both the test plan and the test procedure.

1.1 Goals and objectives

Overall goals and objectives of the test process are described.

1.2 Statement of scope

A description of the scope of software testing is developed. Functionality/features/behavior to be tested is noted. In addition, any functionality/features/behavior that is not to be tested is also noted.

1.3 Major constraints

Any business, product line or technical constraints that will impact the manner in which the software is to be tested are noted here.

2. Test plan

This section describes the overall testing strategy and the project management issues that are required to properly execute effective tests.

2.1 Software to be tested

The software to be tested is identified by name. Exclusions are noted explicitly.

2.2 Testing strategy

The overall strategy for software testing is described.

2.2.1 Unit testing

The unit testing strategy is specified. This includes an indication of the components that will undergo unit tests or the criteria to be used to select components for unit test. Test cases are NOT included here.

2.2.2 Integration testing

The integration testing strategy is specified. This section includes a discussion of the order of integration by software function. Test cases are NOT included here.

2.2.3 Validation testing

The validation testing strategy is specified. This section includes a discussion of the order of validation by software function. Test cases are NOT included here.

2.2.4 High-order testing

The high-order testing strategy is specified. This section includes a discussion of the types of high order tests to be conducted, the responsibility for those tests. Test cases are NOT included here.

2.3 Testing resources and staffing

Specialized testing resources are described and staffing is defined. The role of any ITG is also defined.

2.4 Test work products

The work products produced as a consequence of the testing strategy are identified.

2.5 Test record keeping

Mechanisms for storing and evaluating test results are specified.

2.6 Test metrics

A description of all test metrics to be used during the testing activity is noted here.

2.7 Testing tools and environment

A description of the test environment, including tools, simulators, specialized hardware, test files, and other resources is presented here.

2.8 Test schedule

A detailed schedule for unit, integration, and validation testing as well as high order tests is described.

3. Test Procedure

This section describes as detailed test procedure including test tactics and test cases for the software.

3.1 Software to be tested

The software to be tested is identified by name. Exclusions are noted explicitly.

3.2 Testing procedure

The overall procedure for software testing is described.

3.2.1 Unit test cases

The procedure for unit testing is described for each software component (that will be unit tested) is presented. This section is repeated for all components i.

3.2.1.1 Stubs and/or drivers for component i

3.2.1:2 Test cases component i

3.2.1.3 Purpose of tests for component i

3.2.1.4 Expected results for component i

3.2.2 Integration testing

The integration testing procedure is specified.

3.2.2.1 Testing procedure for integration

3.2.2.2 Stubs and drivers required

3.2.2.3 Test cases and their purpose

3.2.2.4 Expected results

3.2.3 Validation testing

The validation testing procedure is specified.

3.2.3.1　Testing procedure for validation
3.2.3.2　Expected results
3.2.3.3　Pass/fail criterion for all validation tests

3.2.4 High-order testing (a.k.a. system testing)

The high-order testing procedure is specified. For each of the high order tests specified below, the test procedure, test cases, purpose, specialized requirements and pass/fail criteria are specified. It should be noted that not all high-order test methods noted in Sections 3.2.4.n will be conducted for every project.

3.2.4.1　Recovery testing
3.2.4.2　Security testing
3.2.4.3　Stress testing
3.2.4.4　Performance testing
3.2.4.5　Alpha/beta testing
3.2.4.6　Pass/fail criterion for all validation tests

3.3 Testing resources and staffing

Specialized testing resources are described and staffing is defined. The role of any ITG is also defined.

3.4 Test work products

The work products produced as a consequence of the testing procedure are identified.

3.5 Test record keeping and test log

Mechanisms for storing and evaluating test results are specified. The test log is used to maintain a chronological record of all tests and their results.

P.S. This template can be tailored according to the different projects.

A Sample Software Testing Report

Software Testing Specification For Web Publishing System

Version 1.0

Victor Lu (Team leader)
Jack Liang
Jason Wang
Mary Zeng
David Chang

May 19, 2020

Unit 7　Testing the System　系统测试

Table of Contents

1. **Introduction**
   - 1.1　Purpose
   - 1.2　Scope of Testing
   - 1.3　Glossary
   - 1.4　References
   - 1.5　Overview of Document
2. **Verification Testing**
   - 2.1　Unit Relationships
   - 2.2　Unit Testing
   - 2.3　Integrative Testing
3. **System（In-House Validation）Testing**
   - 3.1　Use Case：Search Article
   - 3.2　Use Case：Update Article Status
   - 3.3　Functional Testing Result
      - 3.3.1　Update Author
      - 3.3.2　Update Reviewer
      - 3.3.3　Update Article
      - 3.3.4　Publish Article
4. **Stakeholder Testing**
   - 4.1　Acceptance Testing
   - 4.2　Beta Testing
5. **The Result of Capability Test**
6. **Other Test Results**
   - 6.1　Content Tests
   - 6.2　User Interface Tests
   - 6.3　Safety Requirements
   - 6.4　Portable Requirements
7. **List of Inappropriate Items**
8. **Test Conclusion**

1. Introduction

1.1 Purpose

This document is part of the Software Verification and Validation Plan.

Its primary audiences are the development team, the in-house quality assurance team, and the stakeholders. An important secondary audience is the maintenance team who will need to correct, modify, and enhance this product.

1.2 Scope of Testing

This document will be used by the development team to conduct verification and validation tests for the product. It also details the Acceptance Testing that will be done and the Beta Testing procedures. This document contains only two sample verification tests and two sample use case validation tests in accordance with the class assignment.

1.3 Glossary

Glossary is shown in Table 7-3.

Table 7-3 Glossary

Term	Definition
Acceptance Testing	Testing done in the stakeholder environment. It may be done by the stakeholder, the development team, or an outside team as agreed to by the stakeholder
Beta Testing	The preliminary usage of the system by the stakeholder
Validation Testing	Testing done to assure that the product satisfies all of the needs of the stakeholder and the requirements listed in the Software Requirements Specification
Verification Testing	Testing done to assure that the modules of the project perform as required by the Software Design Description

1.4 References

Victor Lu (Team leader), Jack Liang, Jason Wang, Mary Zeng, David Chang: Software Requirements Specification for Web Publishing System (Version 1.0), Beihang University, 2020.

Victor Lu (Team leader), Jack Liang, Jason Wang, Mary Zeng, David Chang: Software Design Specification for Web Publishing System (Version 1.0), Beihang University, 2020.

1.5 Overview of Document

In the next section, the verification testing about unit and integrated testing for the major design entities including Author Search and Journal Home Page are given.

In the following section, sample validation tests for some of the major use cases as required by a previous assignment are given.

2. Verification Testing

2.1 Unit Relationships

Relationships that do not depend on tables(Table 7-4).

Table 7-4 Relationships Other Than on Tables

Unit Name	Depends on
User (interface)	On-Line Journal (Journal Home Page)
Journal Home Page	AuthorSearch
Journal Home Page	CategorySearch
Journal Home Page	KeywordSearch
Editor (interface)	Article Manager
Article Manager	PeopleManager
PeopleManager	HS Database (interface)
Article Manager	ArtMgr
Article Manager	Publisher

Table dependencies are shown in Table 7-5.

Table 7-5 Table Dependencies

Unit Name	Depends on Table
AuthorSearch	Author
AuthorSearch	Article
CategorySearch	Article
CategorySearch	Category
KeywordSearch	Article
PeopleManager	Reviewer
PeopleManager	ReviewerRelationship
PeopleManager	Author
PeopleManager	AuthorRelationship
ArtMgr	Reviewer
ArtMgr	Author
ArtMgr	AuthorRelationship
ArtMgr	ReviewerRelationship
ArtMgr	CategoryRelationship
ArtMgr	Active Article
Publisher	Active Article
Publisher	Article (Journal Database)

2.2 Unit Testing

Test Unit：AuthorSearch
Xref：SDS 3.1.2 Code：C#
Driver：None needed.

At least five articles must be in the database. The following will suffice (Table 7-6).

Table 7-6 Test Unit: AuthorSearch

Authors	Title	Categories	Abstract	Body
Dean, James Nash, Ogden Paddle, Wheel	Ways to Avoid Testing	Humor, Satire	This article involves ridiculous ideas about mud	< insert block of random text here >
Dean, James	Testing is Fun	Satire	Ridiculous is a manner of speaking	< insert different block of random text here >
Dean, James Double, Mint	Nonsense I Have Known	Romance	No keyword in common	< insert different block of random text here >
Paul, Peter	Apostles of Testing	Software Engineering	Speaking of mud and mud	< insert different block of random text here >
Double, Mint Nash, Ogden	Software is Fun to Create	Software Engineering Romance	Ridiculous mud ideas	< insert different block of random text here >

Method: Use interactive input.

Each author must be tested for each article by that author. Continue each until the article is downloaded. (There are 5 authors to be tested.)

The test is successful when the correct article is downloaded each time.

2.3 Integrative Testing

Test Unit: On-Line Journal (Journal Home Page)

Xref: SDS 3.1.1 Code: C#

Driver: None needed.

Method: This unit is partly tested by testing the Use Case Search Article (see section 3.1 below).

In addition, we must test the E-mail capability. Select to send an E-mail then send a message to the editor of the journal with a text file attachment.

The test is successful if the text file can be successfully downloaded by the editor.

3. System (In-House Validation) Testing

3.1 Use Case: Search Article

Xref: SRS, Section 3.2.1.

Environment: Access the database from a computer on the Internet other than the one where the database resides.

Since this use case uses the Online Journal database exclusively, we will create a copy of that database by entering data directly into tables. The following is a minimum amount of data that must be available for testing:

At least five articles must be in the database. A variety of options to be tested are provided as follows (Table 7-7).

Table 7-7 Test Use Case: Search Article

Authors	Title	Categories	Abstract	Body
Dean, James Nash, Ogden Paddle, Wheel	Ways to Avoid Testing	Humor, Satire	This article involves ridiculous ideas about mud	< insert block of random text here >
Dean, James	Testing is Fun	Satire	Ridiculous is a manner of speaking	< insert different block of random text here >
Dean, James Double, Mint	Nonsense I Have Known	Romance	No keyword in common	< insert different block of random text here >
Paul, Peter	Apostles of Testing	Software Engineering	Speaking of mud and mud	< insert different block of random text here >
Double, Mint Nash, Ogden	Software is Fun to Create	Software Engineering Romance	Ridiculous mud ideas	< insert different block of random text here >

Method: Use interactive input.

There are three functions to be tested, searching by author, category and keyword. They will be tested in random order.

Each author must be tested. Continue each until the article is downloaded.

Each category must be tested. Continue each until the article is downloaded.

Each noun in the Abstract column is to be tested as a keyword. Continue each until the article is downloaded. Try words not contained in the abstract. Make sure that duplicate words do not produce duplicate responses.

Periodically, abandon a search in the middle and restart.

Results Expected: The test is successful if no problems are encountered.

3.2 Use Case: Update Article Status

Xref: SRS, Section 3.2.6

Environment: This test will be performed on the Article Manager Database. The use case Enter Communication must be implemented prior to validation testing of this use case.

At least five articles must be in the database at various status levels. The data listed in Section 2.1 can be used with the addition of status values.

Method: Use interactive input.

There are five functions to be tested, add (or remove) categories, correct typographical errors, remove a reviewer, enter an updated article, and enter a review. After each action is performed, inspect the database to check if the action has occurred.

Add a new category for an article.

Add two new categories for an article.

Delete a category for an article.

Attempt to delete the last category for an article. This should not be allowed.

Modify an author's name.

Modify a title.

Modify an abstract.

Remove a reviewer. If all the other reviews are entered, this should change the article's status accordingly.

Attempt to remove the last reviewer. This should not be allowed.

Enter an updated article.

Enter a review for an article. This should change the reviewer's status accordingly and if that is the last outstanding review, change the article's status accordingly.

Results Expected: The test is successful if no problems are encountered.

3.3 Functional Testing Result

3.3.1 Update Author

Table 7-8 shows the Test Results of Update Author.

Table 7-8 Test Results of Update Author

Test-case No.	101	Function Name	Update Author
Function description	Xref: SRS, Section 2.2.4		
Test steps	Input	New information of the author	
	Output	1. Succeed to update the information of the Author 2. Author Name or ID already exists, and updating fails 3. The non-null field is empty, and updating fails	
Test result	Passed	Problems	None
Tester	Mary Zeng	Test date	05/18/2020

Table 7-9 shows equivalence Partitioning of Update Author.

Table 7-9 Equivalence Partitioning of Update Author

Input Condition	Valid Equivalence Class	No.	Invalid Equivalence Class	No.
Author Name	Author Name is unique and not empty	1	Author Name already exists	3
			Author Name is empty	4
Author ID	Author ID is unique and not empty	2	Author ID already exists	5
			Author ID is empty	6

Table 7-10 shows the test cases of Update Author.

Table 7-10   Test Cases of Update Author

Test-case No.	Input	Expected Output	Coverage
101.1.1	Name：Li Si；ID：1002	Updating succeeded	1,2
101.1.2	Name：Zhang San；ID：1002	Updating failed	3
101.1.3	Name：null；ID：1002	Updating failed	4
101.1.4	Name：Li Si；ID：1001	Updating failed	5
101.1.5	Name：Li Si；ID：null	Updating failed	6

P.S. The record "[Name]：Zhang San [ID]：1001" is in the Author table.

Table 7-11 shows the test result records of Update Author.

Table 7-11   Test Result Records of Update Author

Test-case No.	Expected Result	Actual Result	Test Result	Tester	Test Date
101.1.1	Updating succeeded	Updating succeeded	Passed	Mary Zeng	05/18/2020
101.1.2	Updating failed	Updating failed	Passed	Mary Zeng	05/18/2020
101.1.3	Updating failed	Updating failed	Passed	Mary Zeng	05/18/2020
101.1.4	Updating failed	Updating failed	Passed	Mary Zeng	05/18/2020
101.1.5	Updating failed	Updating failed	Passed	Mary Zeng	05/18/2020

3.3.2 Update Reviewer

Table 7-12 shows the test results of Update Reviewer.

Table 7-12   Test Results of Update Reviewer

Test-case No.	102	Function Name	Update Reviewer
Function description	Xref：SRS，Section 2.2.4		
Test steps	Input	New information of the reviewer	
	Output	1. Succeed to update the information of the reviewer 2. Reviewer Name or ID already exists，and updating failed 3. The non-null field is empty，and updating failed	
Test result	Passed	Problems	None
Tester	Mary Zeng	Test date	05/18/2020

Table 7-13 shows the equivalence partitioning of Update Reviewer.

Table 7-13   Equivalence Partitioning of Update Reviewer

Input Condition	Valid Equivalence Class	No.	Invalid Equivalence Class	No.
Reviewer Name	Reviewer Name is unique and not empty	1	Reviewer Name already exists	3
			Reviewer Name is empty	4
Reviewer ID	Reviewer ID is unique and not empty	2	Reviewer ID already exists	5
			Reviewer ID is empty	6

Table 7-14 shows the test cases of Update Reviewer.

Table 7-14   Test Cases of Update Reviewer

Test-case No.	Input	Expected Output	Coverage
102.1.1	Name: Li Si; ID: 1002	Updating succeeded	1,2
102.1.2	Name: Zhang San; ID: 1002	Updating failed	3
102.1.3	Name: null; ID: 1003	Updating failed	4
102.1.4	Name: Li Si; ID: 1001	Updating failed	5
102.1.5	Name: Li Si; ID: null	Updating failed	6

P.S. The record "[Name]: Zhang San [ID]: 1001" is in the Reviewer table.

Table 7-15 shows the test result records of Update Reviewer.

Table 7-15   Test Result Records of Update Reviewer

Test-case No.	Expected Result	Actual Result	Test Result	Tester	Test Date
102.1.1	Updating succeeded	Updating succeeded	Passed	Mary Zeng	05/18/2020
102.1.2	Updating failed	Updating failed	Passed	Mary Zeng	05/18/2020
102.1.3	Updating failed	Updating failed	Passed	Mary Zeng	05/18/2020
102.1.4	Updating failed	Updating failed	Passed	Mary Zeng	05/18/2020
102.1.5	Updating failed	Updating failed	Passed	Mary Zeng	05/18/2020

3.3.3 Update Article

Table 7-16 shows the test results of Update Article.

Table 7-16   Test Results of Update Article

Test-case No.	103	Function Name	Update Artkle
Function description	Xref: SRS, Section 2.2.4		
Test steps	Input	New information of the article	
	Output	1. Updating succeeded 2. The non-null field is empty, and updating failed 3. Article Name or ID already exists, and updating failed	
Test result	Passed	Problems	None
Tester	Mary Zeng	Test date	05/18/2020

Table 7-17 shows the equivalence partitioning of Update Article.

Table 7-17   Equivalence Partitioning of Update Article

Input Condition	Valid Equivalence Class	No.	Invalid Equivalence Class	No.
Article ID	Article ID is unique and not empty	1	Article ID already exists	7
			Article ID is empty	8
Article Name	Article Name is unique and not empty	2	Article Name already exists	9
			Article Name is empty	10
Article Address	Article Address is not empty	3	Article Address is empty	11
Author Name	Author Name is not empty	4	Author Name is empty	12
Category	Article Category is not empty	5	Article Category is empty	13
Status	Article Status is not empty	6	Article Status is empty	14

Table 7-18 shows the test cases of Update Article.

Table 7-18　Test Cases of Update Article

Test-case No.	Input	Expected Output	Coverage
103.1.1	ID:2; Name:B; Article Address:C:\Users; Author Name:Zhang San; Category:Science; Status:1	Updating succeeded	1,2,3,4,5,6
103.1.2	ID:1; Name:B; Article Address:C:\Users; Author Name:Zhang San; Category:Science; Status:1	Updating failed	7
103.1.3	ID:null; Name:B; Article Address:C:\Users; Author Name:Zhang San; Category:Science; Status:1	Updating failed	8
103.1.4	ID:3; Name:A; Article Address:C:\Users; Author Name:Zhang San; Category:Science; Status:1	Updating failed	9
103.1.5	ID:2; Name:null; Article Address:C:\Users; Author Name:Zhang San; Category:Science; Status:1	Updating failed	10
103.1.6	ID:2; Name:B; Article Address:null; Author Name:Zhang San; Category:Science; Status:1	Updating failed	11
103.1.7	ID:2; Name:B; Article Address:C:\Users; Author Name:null; Category:Science; Status:1	Updating failed	12
103.1.8	ID:2; Name:B; Article Address:C:\Users; Author Name:Zhang San; Category:null; Status:1	Updating failed	13
103.1.9	ID:2; Name:B; Article Address:C:\Users; Author Name:Zhang San; Category:Science; Status:null	Updating failed	14

P.S. The record "[ID]: 1 [Name]: A [Article Address]: C:\Desktop [Author Name]: Zhang San [Category]: Literature [Status]: 1" is in the Article table.

Table 7-19 shows the test result records of Update Article.

Table 7-19　Test Result Records of Update Article

Test-case No.	Expected Result	Actual Result	Test Result	Tester	Test Date
103.1.1	Updating succeeded	Updating succeeded	Passed	Mary Zeng	05/18/2020
103.1.2	Updating failed	Updating failed	Passed	Mary Zeng	05/18/2020
103.1.3	Updating failed	Updating failed	Passed	Mary Zeng	05/18/2020
103.1.4	Updating failed	Updating failed	Passed	Mary Zeng	05/18/2020
103.1.5	Updating failed	Updating failed	Passed	Mary Zeng	05/18/2020
103.1.6	Updating failed	Updating failed	Passed	Mary Zeng	05/18/2020
103.1.7	Updating failed	Updating failed	Passed	Mary Zeng	05/18/2020
103.1.8	Updating failed	Updating failed	Passed	Mary Zeng	05/18/2020
103.1.9	Updating failed	Updating failed	Passed	Mary Zeng	05/18/2020

3.3.4　Publish Article

Table 7-20 shows the test results of Publish Article.

Table 7-20 Test Results of Publish Article

Test-case No.	104	Function Name		Update Artkle
Function description	Xref: SRS, Section 2.2.4			
Test steps	Input	Server IP Address		
	Output	1. Publishing succeeded 2. IP Address is empty, server visiting failed, and publishing failed 3. IP Address is invalid, sever visiting failed, and publishing failed		
Test result	Passed	Problems		None
Tester	Mary Zeng	Test date		05/18/2020

Table 7-21 shows the equivalence partitioning of Publish Article.

Table 7-21 Equivalence Partitioning of Publish Article

Input Condition	Valid Equivalence Class	No.	Invalid Equivalence Class	No.
Server IP Address	IP Address is valid and is not empty	1	IP Address is empty	2
			IP Address is invalid	3

Table 7-22 shows the test cases of Publish Article.

Table 7-22 Test Cases of Publish Article

Test-case No.	Input	Expected Output	Coverage
104.1.1	IP: 172.16.70.245	Publishing succeeded	1
104.1.2	IP: null	Publishing failed	2
104.1.3	IP: 172.16.71.82	Publishing failed	3

P.S. The server IP address is 172.16.70.245.

Table 7-23 shows the test result records of Publish Article.

Table 7-23 Test Result Records of Publish Article

Test-case No.	Expected Result	Actual Result	Test Result	Tester	Test Date
104.1.1	Publishing succeeded	Publishing succeeded	Passed	Mary Zeng	05/18/2020
104.1.2	Publishing failed	Publishing failed	Passed	Mary Zeng	05/18/2020
104.1.3	Publishing failed	Publishing failed	Passed	Mary Zeng	05/18/2020

4. Stakeholder Testing

4.1 Acceptance Testing

The development team will preload the system with at least five articles in various stages of approval based on information provided by the stakeholder. The system will then be installed on the stakeholder's computer. The stakeholder will enter new information and update the preloaded information to test the system. This includes publishing accepted articles to the Online Journal and accessing these from a remote location by multiple simultaneous users.

Any discrepancies between the behavior described in the SRS and the actual system will be corrected and the system will be retested.

4.2 Beta Testing

Any problems detected by the stakeholder within six months of installation will be reported to the development team. After correction, a further six-month period for beta testing will be ensued under the same regulations. All costs for this work are covered by the original contract.

5. The Result of Capability Test

The server-side website can be visited by multiple users without conflict. The local database can be operated by multiple users without throwing any data exceptions.

6. Other Test Results

6.1 Content Tests

The documents (including their texts and figures) contain no typos or grammar mistakes. The final version expected to be delivered to the user has no problems with the structure and contents.

6.2 User Interface Tests

User interface tests may introduce some changes, which we expect with no functional errors or content display problems, and moreover to make the user interface more attractive and user-friendly.

6.3 Safety Requirements

The system should have the capability to prevent itself from being attacked by malicious visits, access, using and revises.

(1) We implement logic level constraints to ensure the correctness of the input data.

(2) We implement the integrity of data while the server is communicating with clients.

(3) We interrupt the communication when the channel is closed in case any bad data sending happens.

6.4 Portable Requirements

This software is Web-based; thus, it is extremely portable.

7. List of Inappropriate Items

There are no inappropriate items.

8. Test Conclusion

Finish Date: 05/19/2020

Test place: Laboratory

Test environment: Typical Home-use Computer

Involved members: Mary Zeng, David Chang, Victor Lu

Advantages of the system: The system is robust and clearly divided by function.

Disadvantages of the system: The functions of the system are not enough and can be further improved.

Inappropriate items: None

 Exercises

1. What is the primary purpose of the tests mentioned in this software test specification?

2. How many functions are to be tested?

3. What dependencies will the test team need in order to fully conduct the tests?

Unit 8

Delivering the System

系统交付

Part 1

Listening & Speaking

Unit 8

Dialogue: Software Deployment

Mr. White:	Welcome to our hotel! I really appreciate[1] your efforts for our Four Seasons Hotel Management Information System.
Kevin:	It's our pleasure. Today, we'll deliver the software to you, including all tested program files, data files and additional documentation for user, such as a user guide and an operator manual. We have also added hypertext "help" files and a troubleshooting guide in our software.
Mr. White:	It sounds wonderful! So can we run the system after deployment right away?
Kevin:	No problem. But I suggest we test the software[2] in a small group of representative users first. After ensuring it runs normally, we can deploy the system in the whole hotel. In this way, those users can first run the software and find problems in practice and the impact will be controlled to a smaller scope and the problems can be solved in time. This is also a common way in software deployment.
Mr. White:	Really? Well, I think it is a good idea. But how much of the scope does it fit? We have 100 staffs and 20 computers in total.
Sharon:	Maybe one third of the information desk in your hotel is ok. In my opinion, information desks deal with most of hotel daily business, and can expose potential problems effectively and efficiently.

[1] Replace with:
1. I am very grateful for…
2. I am very thankful for…

[2] Replace with:
1. I suggest we conduct a beta testing
2. I suggest we conduct a test run

Unit 8　Delivering the System　系统交付

Mr. White：　Ok. I completely agree to this suggestion, and will arrange the equipment and personnel as soon as possible. In addition, I am afraid we would need some timely support from you if some problems are discovered.

Kevin：　Don't worry about it. We will provide installation and start-up assistance, and as a support group, we will ensure troubleshooting assistance for you until the system runs normally for half a year.

Jason：　Here is our contact list with the phone numbers and E-mail addresses. And we need some contact information of your personnel who are in charge of the system in your hotel as well, so as to communicate in a timely manner.

Mr. White：　No problem. I will send you an E-mail listing our contact information within 2 days.

Kevin：　On this aspect, furthermore, I think we had better formally establish problem-logging and feedback mechanisms, which includes what the feedback process is, how you log problems and report them to us, and which form to use, paper and/or electronic, and so on. In this way, we can collect and log your feedback, communicate and assess it with you to determine the modification plan together.

Mr. White：　That's right. Let's go on please.

 Exercises

Work in a group, and make up a similar conversation by replacing the statements with other expressions on the right side.

229

 **Words**

troubleshoot[ˈtrʌblʃuːt] v. 解决重大问题,排除……的故障
timely[ˈtaimli] adj. 及时的,适时的
start-up 启动阶段的,开始阶段的
log[lɔg] v. 记录

 Phrases

be grateful for 对……心存感激
on this aspect 在这个方面

Listening Comprehension: Software Delivery

Listen to the article and answer the following 3 questions based on it. After you hear a question, there will be a break of 15 seconds. During the break, you will decide which one is the best answer among the four choices marked (A), (B), (C) and (D).

Questions

1. Which is the most appropriate order of steps in the software delivery stage?
   (A) Fixing, delivery, feedback, support, and modification
   (B) Delivery, fixing, support, feedback, and modification
   (C) Fixing, support, delivery, feedback, and modification
   (D) Delivery, fixing, feedback, support, and modification

2. Which point is not correct on support before the software delivery according to this article?
   (A) Support should be considered after the software delivery because the possible troubles are unpredictable during the practical running.
   (B) The software team can conduct a categorical assessment of the kinds of support requested by an appropriate record keeping mechanisms.
   (C) Support documents should be provided on the software delivery.
   (D) A support regime must be established before the software is delivered.

3. Which point is correct on feedback after the software delivery according to this article?
   (A) Feedback should be responded and resolved in the present delivered increment immediately and entirely.
   (B) Feedback should not be answered by the software team anymore because the

project has finished since the software delivery.
(C) Feedback allows the customer to suggest changes that have business value and provide the input for the next iterative software engineering cycle for the software team.
(D) Feedback is undesirable because of its resulting in some uncontrolled requirements and unexpected changes.

Words

cheer[tʃɪə(r)] n. 欢呼，喝彩	按类别的
regime[reɪˈʒiːm] n. 体制，体系	accommodate[əˈkɒmədeɪt] v. 考虑到，顾及
categorical[ˌkætəˈɡɒrɪkl] adj. 分类的，	

Phrases

put into use 应用，使用

Dictation: Bug and Debugging

This article will be played three times. Listen carefully, and fill in the numbered spaces with the appropriate words you have heard.

The word "bug" has been used to describe a "**bogeyman**" ever since the fourteenth century. In computers, it **was** ___1___ attributed to **Admiral** Dr. Grace Hopper, the inventor of COBOL, who ___2___ the first computer bugs in the 1940s. **Albeit** not the flying kind, software ___3___ **manifest** themselves in a variety of ways, from misunderstood requirements to coding ___4___.

Having been used as a term in **aeronautics** before ___5___ the world of computers, the word "debugging" is first used in three papers from 1952 **ACM** National Meetings. A software engineer, evaluating the results of a test, **is** often **confronted with** a "**symptomatic**" ___6___ of a software problem. That is, the ___7___ manifestation of the error and the ___8___ cause of the error may have no obvious ___9___ to one another. The poorly understood **mental** process that ___10___ a symptom to a cause is debugging. In detail, in computers, debugging is the process of ___11___ and fixing bugs in computer program code or the engineering of a hardware ___12___. It is a necessary process in almost any new software or hardware development process, whether a ___13___ product or an enterprise or personal ___14___ program.

No one writes perfect software, so it's a **given** that debugging will **take up** a major

15   of your days. In his book Debugging, David, J Agans provides the 9 Indispensable Rules for Finding Even the Most Elusive Software and Hardware Problems as follow: Understand the system, make it   16   ,   17   thinking and look,   18   and conquer, change one thing at a time, keep an audit   19   , check the plug, get a   20   view, and if you didn't fix it, it ain't fixed.

Words

bogeyman[ˈbəʊɡɪmæn] n. 可怕(或可憎)的人或物, 妖怪, 精灵	表明, 表示
admiral[ˈædmərəl] n. 海军上将, 海军将官	mental[ˈmentl] adj. 思考的, 智力的, 脑力的
albeit[ɔːlˈbiːɪt] conj. 虽然	given[ˈɡɪvn] n. 假设
manifest[ˈmænɪfest] v. 显示, 表明	elusive[ɪˈluːsɪv] adj. 难以被发现的
aeronautics[ˌeərəˈnɔːtɪks] n. 航空学	audit[ˈɔːdɪt] n. (质量或标准的)审查, 检查
symptomatic[ˌsɪmptəˈmætɪk] adj. 有症状的	plug[plʌɡ] n. 插头, 插座
manifestation[ˌmænɪfeˈsteɪʃn] n. 显示,	ain't[eɪnt] (= are not, am not, is not) 不是

Phrases

be attributed to　　归因于
be confronted with　　面对
take up　　占(空间、时间)

Abbreviations

ACM　　Association for Computing Machinery　　国际计算机学会

Part 2

Reading & Translating

Section A: Software Maintenance

　　System development is complete when the system is operational, that is, when the system is being used by users in an actual production environment. Any work done to change the system after it is operational is considered to be maintenance. Many people think of software system maintenance as they do hardware maintenance: repair or prevention of broken or improperly working parts. However, software maintenance cannot be viewed in the same way (Figure 8-1). Let us see why.

Figure 8-1　Software Maintenance is Different from Hardware Maintenance

One goal of software engineering is developing techniques that define a problem exactly, design a system as a solution, implement a correct and efficient set of programs, and test the system for faults. This goal is similar for hardware developers: producing a reliable, fault-free product that works according to specification. Hardware maintenance in such a system concentrates on replacing parts that wear out or using techniques that prolong the system's life. However, while do constructs do not wear out after 10,000 loops, and semicolons do not fall off the ends of statements. Unlike hardware, software does not degrade or require periodic maintenance. Thus, software systems are different from hardware, and we cannot use hardware analogies successfully the way we can for other aspects of software engineering.

The biggest difference between hardware and software systems is that software systems are built to incorporate change. Except for the simplest cases, the systems we develop are evolutionary. That is, one or more of the system's defining characteristics usually change during the life of the system.

Software systems may change not just because a customer makes a decision to do something in a different way, but also because the nature of the system itself changes. For example, consider a system that computes payroll deductions and issues paychecks for a company. The system is dependent on the tax laws and regulations of the city, state or province, and country in which the company is located. If the tax laws change, or if the company moves to another location, the system may require modification. Thus, system changes may be required even if the system has been working acceptably in the past.

When we develop systems, our main focus is on producing code that implements the requirements and works correctly. At each stage of development, our team continually refers to work produced at earlier stages. The design components are tied to the requirements specification, the code components are cross-referenced and reviewed for compliance with the design, and the tests are based on finding out whether functions and constrains are working according to the requirements and design. Thus, development involves looking back in a careful, controlled way.

Maintenance is different. As **maintainers**, we look back at development products, but also at the present by establishing a working relationship with users and operators to find out how satisfied they are with the way the systems work. [1] We look forward, too, to **anticipate** things that might go wrong, to consider functional changes required by a changing business need, and to consider system changes required by changing hardware, software, or interfaces. Thus, maintenance has a broader scope, with more to track and control. Let us examine the activities needed to keep a system running **smoothly** and identify who performs them.

Maintenance activities are similar to those of development: analyzing requirements, evaluating system and program design, writing and reviewing code, testing changes, and updating documentation. So the people who perform maintenance—analysts, designers, and programmers—have similar roles. However, because changes often require an **intimate** knowledge of the code's structure and content, programmers play a much larger role in maintenance than they did in development.

Maintenance focuses on four major aspects of system evolution simultaneously:

(1) maintaining control over the system's day-to-day functions

(2) maintaining control over system modifications

(3) **perfecting** existing acceptable functions

(4) preventing system performance from degrading our unacceptable levels

Often, a separate group of analysts, designers, and programmers (sometimes including one or two members of the development team) **is designated as** the maintenance team. A fresh, new team may be more objective than the original developers.

Maintaining a system involves all team members. Typically, users, operators, or customer representatives **approach** the maintenance team with a comment or problem. The analysts or programmers determine which parts of the code are affected, the impact on the design, and the likely resources (including time and effort) to make the necessary changes. The team **is involved in** many activities:

(1) understanding the system

(2) locating information in system documentation

(3) keeping system documentation up to date

(4) extending existing functions to accommodate new or changing requirements

(5) adding new functions to the system

(6) finding the source of system failures or problems

(7) locating and correcting faults

(8) answering questions about the way the system works

(9) restructuring design and code components

(10) rewriting design and code components

(11) deleting design and code components that are no longer useful
(12) managing changes to the system as they are made

In addition, maintenance team members work with users, operators, and customers. First, they try to understand the problem as expressed in the user's language. Then, the problem is transformed into a request for modification. The change request includes a description of how the system works now, how the user wants the system to work, and what modifications are needed to produce the changes. Once design or code is modified and tested, the maintenance team retrains the user, if necessary. Thus, maintenance involves interaction with people as well as with the software and hardware.

Words

loop[lu:p] n. 循环 semicolon[ˌsemiˈkəulən] n. 分号 degrade[diˈgreid] v. 退化, 削弱(尤指质量) periodic[ˌpiəriˈɔdik] adj. 定期的, 周期的 analogy[əˈnælədʒi] n. 类比, 类推 incorporate[inˈkɔ:pəreit] v. 加入, 包含 evolutionary[ˌi:vəˈlu:ʃənri] adj. 逐渐发展的, 演变的 cross-reference 前后对照, (书等) 加注使相互参照	maintainer[meinˈteinə] n. 软件维护人员 anticipate[ænˈtisipeit] v. 预料, 预见, 预计(并做准备) smoothly[ˈsmu:ðli] adv. 平稳地, 连续而流畅地, 顺利地 intimate[ˈintimət, ˈintimeit] adj. 精通的, 基本的 perfect[ˈpə:fikt, pəˈfekt] v. 使完善, 改进 approach[əˈprəutʃ] v. 接洽

Phrases

wear out   用坏, 磨损
fall off   脱落, 掉队
except for   除……之外
be tied to   被束缚
for compliance with   遵守, 遵从, 符合
be designated as   被指定为
be involved in   参与

Notes

[1]　**Original**: As maintainers, we look back at development products, but also at

the present by establishing a working relationship with users and operators to find out how satisfied they are with the way the systems work.

Translation：作为软件维护人员，我们不但要回头看开发的产品，而且要看眼前的工作，通过与用户和操作人员建立一种工作关系，以发现他们对于系统工作方式的满意程度。

 Exercises

Ⅰ. Read the following statements carefully，and decide whether they are true（T）or false（F）according to the text.

    ____ 1. Maintenance begins with any work done to change the system after it is operational.

    ____ 2. Software maintenance is similar to hardware maintenance.

    ____ 3. Maintenance activities are similar to those of development.

    ____ 4. Often，a maintenance team involves analysts，designers，and programmers.

    ____ 5. Maintenance requires the cooperation with users，operators，and customers.

Ⅱ. Choose the best answer to each of the following questions according to the text.

1. Which of the following does not pertain to the characteristics of software maintenance?

    (A) It does not require to be performed periodically.

    (B) It has a broader scope，with more to track and control than development.

    (C) It concentrates on replacing parts that wear out or using techniques that prolong the system's life.

    (D) It involves interaction with people as well as with the software and hardware.

2. Which of the following is wrong about the change of software?

    (A) It derives from customers' new decisions to do something in different ways.

    (B) A changing business need may bring some functional changes of software.

    (C) A changing hardware，software，or interfaces need may bring some system changes of software.

    (D) It can be resulted from the change of the natural of the system itself.

3. Which of the following is wrong about the people who perform maintenance?

    (A) Users，operators，or customer representatives are also involved although they do not belong to the maintenance team.

    (B) Programmers play a much larger role in maintenance than they did in development.

    (C) The original group of programmers is more appropriate for maintenance than a fresh, new team.

(D) Maintainers must look back, at the present, but also forward.

Ⅲ. Fill in the numbered spaces with the words or phrases chosen from the box. Change the forms where necessary.

> besides   whether   on   case   as
> first   unfortunate   mean   specific   involve

Understanding Software Release Management

  ___1___ it is dictated by competition, by technology, or by customer demands, the release of new software, or upgrades, is an integral part of the ever-evolving world of Information Technology. Software Release Management ___2___ the application of Project Management Principles to the deployment of new software packages, or upgrades to existing packages.

  ___3___ the process areas defined in A Guide to the Project Management Body of Knowledge (PMBOK® Guide), namely: Initiating, Planning, Executing, Controlling and Closing; the deployment of software involves the application of special processes ___4___ to Release Management.

  They include: Functional Product Request Process, Release Packaging Process, Documentation Process, Development Process, Change Control Process, Customer Testing Process, Customer Notification Process, Training Process, and Deployment Process.

  Also included ___5___ part of Release Management is the management of the usual Project Management knowledge areas of Scope, Time, Cost, Risk, Contract, Human Resources, Communication, and Quality.

  ___6___, until very recently, IT companies did not have Release Managers. In most ___7___, the Program Manager would "keep an eye" on the activities of the various teams. This simply ___8___ keeping track of the deployment schedule. The tide has begun to change, and a few companies are now seeing the need for Release Managers.

  The Release Manager is ___9___ and foremost a Project Manager whose job is to manage the release of the software from conception to deployment. He is not just a schedule administrator; he is a negotiator, a coordinator, a communicator, and at times a mediator. He is proactive, and is aware of the activities of all the stakeholders and the impact of these activities on the ultimate objective: the deployment of a quality product on schedule and ___10___ budget.

  He is the leader of the cross-functional team that includes all the stakeholders.

Ⅳ. **Translate the following passage into Chinese.**

Types of Software Maintenance

There are four types of maintenance, namely, corrective, adaptive, perfective, and preventive. Corrective maintenance is concerned with fixing errors that are observed when the software is in use. Adaptive maintenance is concerned with the change in the software that takes place to make the software adaptable to new environment such as to run the software on a new operating system. Perfective maintenance is concerned with the change in the software that occurs while adding new functionalities in the software. Preventive maintenance involves implementing changes to prevent the occurrence of errors. The distribution of types of maintenance is determined by type and by percentage of time consumed.

Section B: Why Software Delivery Management Matters

Software delivery has come a long way in the last 10 years. Many organizations have scrapped the restrictive waterfall model in favor of collaborative approaches that enable them to build faster, change features midstream and deliver updates continuously. But today's delivery processes aren't as efficient as they could be. They're still fraught with bottlenecks, mistakes and confusion. Teams are using too many tools that perform specific functions and don't tie together well. Data is scattered throughout departments with no single pane of glass [1] to provide visibility into the entire delivery life cycle. As much as we've moved forward, we're not there yet.

There's a fundamental disconnect in the way companies continue to deliver software. Creating DevOps [2] cultures and adopting continuous delivery practices have helped, but companies still aren't managing with the precision necessary to meet the demands placed on industry in the future. What's needed is a more holistic approach to the management of software delivery.

That approach is outlined in a new model called Software Delivery Management (SDM) (Figure 8-2). It's more than just an industry term. It's a revolutionary new framework that outlines ways to remove the blockers that lead to inefficiencies—to connect the people, tools and processes that play key roles in software delivery.

In an enterprise setting, there are often a variety of development methods, tools and technology stacks being used to deliver a wide range of software to meet different needs, through different processes. SDM doesn't replace DevOps, Continuous Integration or Continuous Delivery (CI/CD) [3]. These are bedrock concepts that will be essential to software delivery for years to come. Instead, it builds on them.

Even in those companies with a mature CI/CD pipeline, multiple daily deployments and a full, company-wide commitment to DevOps, there is often no end-to-end insight into the value stream — where products and features are stuck now or get stuck frequently, where bottlenecks and inefficiencies slow down value delivery to end users.

Unit 8　Delivering the System 系统交付

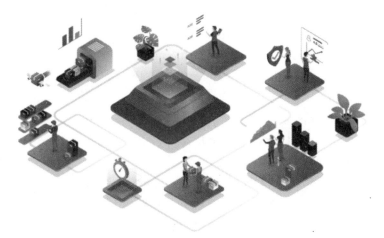

Figure 8-2　Software Delivery Management

Teams following DevOps "best practices" often do things completely differently from each other, even within the same company. There is also a complete inability to understand how software affects business Key Performance Indicators (KPIs) [4].

CI/CD helps teams accelerate software delivery — but it doesn't ensure that they are delivering the right software, or that the business need is being met. It doesn't pull together data and artifacts across the entire software delivery life cycle from the many siloed tools an organization relies on, to provide a single overview with the contextual information that developers, product managers, operations teams, product marketers and support teams need. CI/CD doesn't provide the data needed to measure how well the software organization is creating value for the business — and without a way to measure, software organizations have no way to know if they are reaching prescribed KPIs or even improving.

Software Delivery Management evolves CI/CD in several ways. It proposes the use of connected tools, so stakeholders aren't working in technology silos [5]. It proposes a holistic data model, so all stakeholders in the organization are accessing and sharing the same information. It also extends the feedback loop to encompass the entire application lifecycle, from issue creation to end users interacting with the application.

Just like DevOps breaks down the walls between the development and operations teams, SDM breaks down the walls to connect software delivery to cross functional teams across the organization. It allows them to communicate and collaborate better, through a unified process to ultimately make better software faster, that also effectively addresses the business needs and creates value for the customer.

Development and operations teams talk a lot about speed and frequency: if we can deliver hundreds or thousands of times per day, we have a cutting-edge software development process. At the end of the day, though, software delivery is not just about speed: it is about delivering the best possible product — one that meets customers' needs

— as quickly as possible. Speeding up delivery of poor quality software or software that doesn't address the market need accomplishes nothing.

While SDM still places value on speed, the emphasis is on value delivery across the entire software organization through application of the four pillars of SDM:

(1) Common data

(2) Universal insights

(3) Common connected processes and

(4) All functions collaborating

The goal isn't just to automate everything as a way to deliver instantaneously, but rather to free the creative humans who build software from low-value tasks and allow them to spend more time and energy on high-value, collaborative and creative work. And to give the entire organization visibility into the data required to make software drive better business outcomes.

In contrast to stereotypes about the lone, genius developer, companies get the highest value out of collaborative, cross-functional efforts. A DevOps culture is a step in the right direction, but true cross-functionality should involve collaboration among all stakeholders in the software organization, all the way through to customer success. When this happens, everyone has more time to focus on high-value work, from developers who no longer blindly create features to customer support specialists who are empowered to fix the root cause of customer frustration instead of repeatedly addressing the same customer concerns.

Adopting SDM means not having to sort through customer data to deduce what the problem statement might be, then having non-developers create the requirements for a new feature to address those problems — then handing the requirement set to the software organization without ever discussing the original data.

Instead, customer support teams, product stakeholders and developers have access to the same shared database and same information about customer usage. They can work together to both identify the problem and brainstorm how software could be used to solve it. Everyone understands the trade-offs different potential solutions involve, and can decide collectively how best to prioritize and plan feature development to get the desired business results, while delivering value to the customer.

It's more likely the developers will deliver the right product the first time — and if they don't, they have enough information in the feedback loop to iterate until the software solves the customer need. Everyone, from developers to product marketers to support teams, has visibility into the process and knows when to expect new features or updates.

Software delivery has come a long way. Companies are building faster, updating more often and releasing higher-quality applications than ever before. But, as technologies advance and business sectors evolve, enterprises will face pressure to do more and do it better. There's still a lot of room for improvement. The industry can get there, if it approaches software delivery holistically and defines it as a core business function.

Words

scrap [skræp] v. 废弃，取消
pane [pein] n. （一片）窗玻璃
holistic [hə'listik] adj. 整体的，全面的，功能整体性的
stack [stæk] n. 一堆，大量，许多
bedrock ['bedrɔk] n. 牢固基础，基本事实
commitment [kə'mitmənt] n. 提交，承诺
silo ['sailəu] v & n. 将……贮藏在筒仓中，筒仓，数据仓库
contextual [kən'tekstʃuəl] adj. 与上下文有关的，与语境相关的

prescribe [pri'skrab] v. 规定，命令，指示
cutting-edge 先进的
instantaneous [ˌinstən'teiniəs] adj. 立即的，立刻的，瞬间的
stereotype ['steriətap] n. 模式化观念（或形象），老一套，刻板印象
genius ['dʒi:niəs] n. 天才，天才人物
cross-functional 跨职能的
brainstorm ['breinstɔ:m] v. 头脑风暴，集中各人智慧猛攻

Phrases

come a long way 突飞猛进，取得很大进展
in favor of 支持，赞同
features midstream 中游特色
be fraught with 充满，充满……的
break down 拆除，消除，破除（障碍或偏见）
through to 直到，一直到

Notes

[1] 单一的玻璃窗格（a single pane of glass），IT 术语，可以等同于仪表板，其语义来自"单块玻璃或单格的窗户"，一般指 IT 资源监控和管理。

[2] DevOps（Development 和 Operations 的组合词）是一组过程、方法与系统的统称，用于促进开发（应用程序/软件工程）、技术运营和质量保障（QA）部门之间的沟通、协作与整合。

它是一种重视"软件开发人员（Dev）"和"IT 运维技术人员（Ops）"之间沟通合作的文化、运动或惯例。通过自动化"软件交付"和"架构变更"的流程，来使得构建、测试、发布软件能够更加地快捷、频繁和可靠。

241

它的出现是由于软件行业日益清晰地认识到：为了按时交付软件产品和服务，开发和运维工作必须紧密合作。

［3］（1）持续集成（CI）是在源代码变更后自动检测、拉取、构建和（在大多数情况下）进行单元测试的过程。持续集成是启动管道的环节。

持续集成的目标是快速确保开发人员新提交的变更是好的，并且适合在代码库中进一步使用。

（2）持续交付（CD）通常是指整个流程链（管道），它自动监测源代码变更并通过构建、测试、打包和相关操作运行它们以生成可部署的版本，基本上没有任何人为干预。

持续交付在软件开发过程中的目标是自动化、效率、可靠性、可重复性和质量保障（通过持续测试）。

持续交付包含持续集成（自动检测源代码变更、执行构建过程、运行单元测试以验证变更）、持续测试（对代码运行各种测试以保障代码质量）和（可选）持续部署（通过管道发布版本自动提供给用户）。

［4］关键绩效指标（Key Performance Indicator，KPI）是通过对组织内部流程的输入端、输出端的关键参数进行设置、取样、计算、分析，衡量流程绩效的一种目标式量化管理指标，是把企业的战略目标分解为可操作的工作目标的工具，是企业绩效管理的基础。KPI可以是部门主管明确部门的主要责任，并以此为基础，明确部门人员的业绩衡量指标。建立明确的切实可行的KPI体系，是做好绩效管理的关键。关键绩效指标是用于衡量工作人员工作绩效表现的量化指标，是绩效计划的重要组成部分。

［5］筒仓（silo）是指企业内部因缺少沟通，部门间各自为政，只有垂直的指挥系统，没有水平的协同机制，就像一个个的谷仓，拥有各自独立的进出系统，但缺少了谷仓与谷仓之间的沟通和互动。

Exercises

Ⅰ. Read the following statements carefully, and decide whether they are true（T）or false（F）according to the text.

　　____ 1. SDM proposes the use of connected tools, so stakeholders aren't working in technology silos.

　　____ 2. SDM can replace DevOps, Continuous Integration or Continuous Delivery.

　　____ 3. CI is an abbreviation of Computer Integration.

　　____ 4. CD is an abbreviation of Computer Delivery.

　　____ 5. SDM breaks down the walls to connect software delivery to cross functional teams across the organization.

Ⅱ. Choose the best answer to each of the following questions according to the text.

　　1. Which of the following is right about SDM?

　　　　（A）SDM breaks down the walls to connect software delivery to cross functional teams across the organization.

(B) Adopting SDM means not having to sort through customer data to deduce what the problem statement might be.

(C) SDM is a revolutionary new framework that outlines ways to remove the blockers that lead to inefficiencies—to connect the people, tools and processes that play key roles in software delivery.

(D) All of the above

2. Which of the following is wrong about DevOps?

(A) DevOps breaks down the walls between the development and operations teams.

(B) DevOps can replace SDM.

(C) Teams following DevOps "best practices" often do things completely differently from each other, even within the same company.

(D) None of the above.

3. Which of the following is wrong about CI/CD?

(A) CI/CD doesn't provide the data needed to measure how well the software organization is creating value for the business.

(B) CI/CD helps teams accelerate software delivery.

(C) CI/CD doesn't ensure that teams are delivering the right software, or that the business need is being met.

(D) None of the above.

Ⅲ. Fill in the numbered spaces with the words or phrases chosen from the box. Change the forms where necessary.

> deal   match   know   access   depend
> term   do   include   respect   see

Types of Software Quality

There are different types of software quality which ___1___ on the contexts they are used in software engineering and maintenance. One of the most basic types of software quality is software functional quality. The functional quality is ___2___ as the ways it conforms or complies to a specific design based on functional specifications or requirements. This description can equally be ___3___ as its compatibility with its intended use or how competitive the software is in the marketplace with ___4___ to its importance.

The structural quality is also a type of software quality. It ___5___ with the nonfunctional needs that support and help the delivery of its functional requirements.

These nonfunctional requirements ___6___ maintainability, robustness, and the level at which the software was manufactured correctly. These types of software quality can be ___7___ through the appraisal of the software source code and its inner structure at the system level, the technology level, and the unit level. The measure and analysis of these types of software quality is ___8___ through software testing.

The main obstacle in effectively defining software quality is the inability to ___9___ the future needs of the user with the present features of the products. The constant change in consumer needs and the competition inherent in the software market makes it quite difficult to always provide all the software needs of the end user in ___10___ of quality.

Ⅳ. **Translate the following passage into Chinese.**

Quality in Software Engineering

In broader terms, the software quality definition of "fitness for purpose" refers to the satisfaction of requirements. But what are requirements? Requirements, also called user stories in today's Agile terms, can be categorized as functional and nonfunctional. Functional requirements refer to specific functions that the software should be able to perform. For example, the ability to print on an HP M1136 printer is a functional requirement. However, just because the software has a certain function or a user can complete a task using the software, does not mean that the software is of good quality. There are probably many instances where you've used software and it did what it was supposed to do, such as find you a flight or make a hotel reservation, but you thought it was poor quality. This is because of "how" the function was implemented. The dissatisfaction with "how" represents the nonfunctional requirements not being met.

Part 3

Simulated Writing: User Guide

User guides are written to explain in layman's terms how to use software. Generally, they accompany other documentation, such as system administration guides. They are generally the first part of call when something needs to be read. People read them in a hurry, usually when they are frustrated and losing patience with the software.

As the name implies, user guides are written for users to help them understand an application or system. When writing a user guide, use simple language with short sentences. This style helps the user through the application. User guides are often written for non-technical individuals. The level of content and terminology will differ considerably from for example a system administration guide, which is very detailed and complex.

Identifying Your Audience

Before you start, identify the target audience, their level of technical knowledge, and how they will use the guide. In the planning process, develop an audience definition that identifies the user, the system and the tasks.

Software is used to do specific things. Users want to know what the software can do for them, for example, how to print a page in landscape. The user guide is to teach them how the software helps them to do something.

Writing the Parts of a User Guide

The following section describes special considerations for the front matter, body, and back matter of the guide.

Front Matter

Include a cover page, table of contents and maybe a preface.

- **Cover and Title Page.** If the document is copyrighted, include a copyright notice: Copyright © 2020 The Name of Your Company. Place this on the cover (and also the title page).
- **Disclaimer.** Include a standard disclaimer inside the front cover.
- **Preface.** Use this section to reference other documents related to the software. It is common practice to include a section on "How to Use This Manual" as an introduction.
- **Contents.** Always have a table of contents, and include an index at the end of the document.

Body of the Guide

In the main body, distinguish and separate procedures from reference materials.

- **Procedures.** Procedures help the user perform specific tasks and may include:

(1) When, why, and how to perform a task

(2) What the screen will show after performing a task

(3) Examples of tasks and program operation.

Procedures involve:

(1) Listing the major tasks.

(2) Breaking each major task into subtasks.

(3) Leading the user through each subtask in a series of steps.

(4) Using an "if-then" approach in explaining decisions that users can make.

- **Learning to chunk.** Breaking tasks and information into smaller parts is called "chunking."

In user guides, information can be divided by menu options and the consequences. Subject matter and subtasks that need to be performed divide chunks. Each chunk can form a new chapter or section of a manual. When writing procedures, number each step. If you only need a few steps, set them apart with white space. Use a consistently structured format for sections. For instance, you could

(1) Begin each section with an overview of the activities to perform

(2) Describe the input and the output

(3) Provide the procedures for accomplishing the tasks.

- **Using the if-then approach.** When users must make choices, use an if-then approach to provide clear alternatives and consequences. And use diagrams for complicated procedures.
- **Reference materials.** Readers study reference materials when they need more information on a particular topic. Reference materials can include:

(1) Program options.

(2) Dictionary of "hot" keys.

(3) Error messages.

(4) Troubleshooting tables.

(5) Commonly asked questions.

(6) Examples of input and output.

Back Matter

Glossaries and indexes (normally placed behind the procedures) are two of the most valuable accessories in user's guides.

- **Glossary.** This is vital when you create new, program-specific terms or when you use common terms in a special way.

    (1) A small glossary can appear at the front before the contents pages

    (2) A larger complete glossary should appear in the back matter.

    (3) Glossary terms should be flagged the first time they appear in text.

- **Index.** Any guide longer than 20 pages benefits from an index. Large documents without an index are impossible to use efficiently.

Establishing Standards

- **Format**

Some clients may have format preferences. Check this with the client during planning. A good format maps the organization of the guide, using clear signposts to find reference information easily.

To achieve this, use the following tactics:

(1) Use headings for organizing information.

(2) Place page numbers and section titles on every page, either in footers or headers.

(3) Use section numbering with restraint. As an alternative to section numbering, divide instructions into manageable segments, based on activities that the reader will

Unit 8 Delivering the System 系统交付

perform. Provide new sections on the right-hand pages for each new division.

(4) Dual columns can be used effectively by putting headings in the left-hand column and the text in the right-hand column.

- **Style**

Use an appropriate style. Decide on the technical level of your language, how you address the user, and conventions that are required.

- **Technical Language**

Match the level of technical language with the audience's level of proficiency. Always underestimate the knowledge of your readers rather than overestimate it.

Limit technical terms to those the user will encounter. If you must define a large number of terms, use a glossary to supplement definitions in the text.

- **Addressing the User**

When writing procedures, use the active voice (e. g. Click this) and address users directly (write "you" rather than "the user"). When explaining an action, use the "command" form of the verb.

P. S. This guide can be tailored according to the different user guides.

A Sample User Guide

User Guide for 7-Zip

How to use 7-Zip to archive and compress your files? 7-Zip is a great file compression and archiving tool. However, it's unfortunate how only professionals and advanced users are the ones who know about it. So to give credit to this amazing tool, we will teach you a simple guide on how to use 7-Zip.

There are too many file compression programs available online and among the most commonly used nowadays are WinZip and WinRAR. On the other hand, a lesser-known choice is 7-Zip. It is free and has the smallest compression compared to the prior options.

The thing is, not many are aware that it is available till the time that it is introduced to them. Thus, they are often clueless about how to use 7-Zip or the benefits it has to offer. They are also not aware that it can also have archive errors sometimes.

Is 7-Zip safe? Which is a better tool, 7-Zip vs WinRAR? Those are the few questions you need to learn after exploring how to use 7-Zip so you can manage the 7-Zip command line.

What is 7-Zip?

Well, 7-Zip as mentioned earlier, is a file archiver coupled with a high compression ratio both for GZIP and ZIP formats. This is between 2 to 10 percent better than other tools available.

7-Zip boosts its own format delivering a high compression ratio that is roughly 40 percent higher. This is primarily because 7-Zip is using LZMA as well as LZMA compression. In addition, it has dictionary sizes and superb compression settings.

Zip tool has gained its appeal and worldwide interests primarily because of its ability to compress files efficiently. However, this doesn't stop 7-Zip to show that it can match the bigger and more established names in the industry.

How to Install 7-Zip

After the 7-Zip download and launching, you will be amazed by how easy and simple to navigate through its user interface. In downloading 7-Zip for Mac or Linux, just follow the steps discussed.

1. Go to 7-Zip's home page and choose the version you wish to download (Figure 8-3).

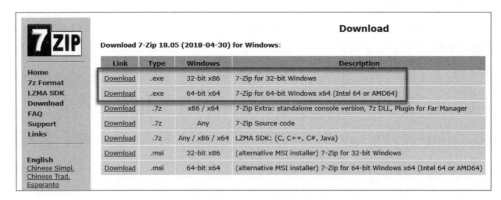

Figure 8-3　7-Zip's Home Page

2. As of this writing, these are the versions available.

3. Choose the version you want and click on the **"Download"** button.

4. A new window appears which asks you what directory you like to install the program. Choose the destination folder you wish to save the program and click **"Install"** (Figure 8-4).

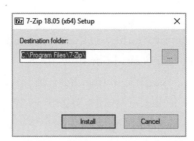

Figure 8-4　Choose the Destination Folder You Wish to Save the Program

5. Once the 7-Zip is installed, hit the **"Finish"** button.

That's everything you have to do. Now, let's proceed on how to use 7-Zip and other vital product details and information.

Navigating through 7-Zip's UI

Inside the window, you'll see the main toolbar containing the most useful features and other menus that let you dig deeper and be accustomed to its UI. Some of the features are briefly discussed below:

- **Extract**—a button that allows you to accept or browse for the default destination path for your file easily.
- **View**—this menu contains the "Folder History" as well as the "Favorites" menu which allows you to save to as much as ten folders.

7-Zip can integrate with the Windows Explorer menu which displays archive files as folders. At the same time, it provides a toolbar with a drag-and-drop feature. It is possible as well to switch between a single or even dual-pane view. 7-Zip password protect can help you protect the program.

How to Use 7-Zip: Compression

We've discussed and learned the installation procedure as well as a quick overview for 7-Zip. Now, we will move on to properly use 7-Zip both for compressing files and extracting them. At first, let's proceed with compressing files.

- **Step number 1.** Open 7-Zip File Manager (Figure 8-5).

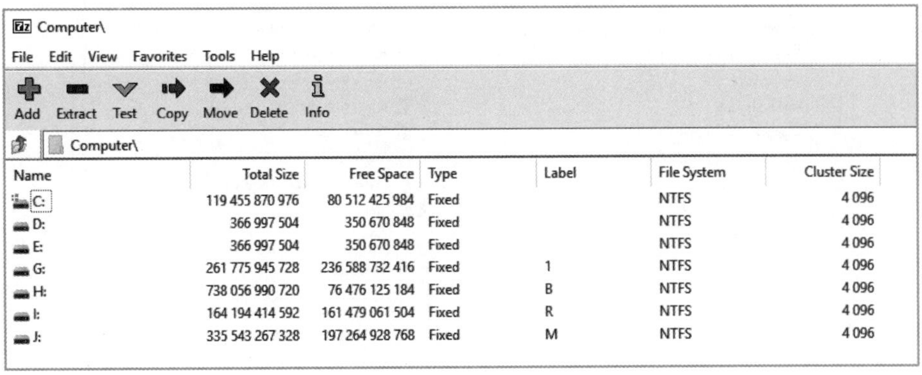
Computer\
File Edit View Favorites Tools Help
Add Extract Test Copy Move Delete Info
Computer\
Name Total Size Free Space Type Label File System Cluster Size
C: 119 455 870 976 80 512 425 984 Fixed NTFS 4 096
D: 366 997 504 350 670 848 Fixed NTFS 4 096
E: 366 997 504 350 670 848 Fixed NTFS 4 096
G: 261 775 945 728 236 588 732 416 Fixed 1 NTFS 4 096
H: 738 056 990 720 76 476 125 184 Fixed B NTFS 4 096
I: 164 194 414 592 161 479 061 504 Fixed R NTFS 4 096
J: 335 543 267 328 197 264 928 768 Fixed M NTFS 4 096

Figure 8-5 7-Zip File Manager

- **Step number 2.** Click on what file you wish to compress and click "**Add**". This will identify the file that you need to compress.
- **Step number 3.** "**Add to Archive**" window is going to pop up (Figure 8-6). Ensure that your Archive format is set to "**Zip**" and then hit the "**OK**" button.
- **Step number 4.** 7-Zip will now process the file and compress it in a zip file that's located in the same destination where the original file is.

Extraction

When extracting files, there are 5 straightforward steps you need to do. Just check the instructions below.

1. Select the files that you want to extract.

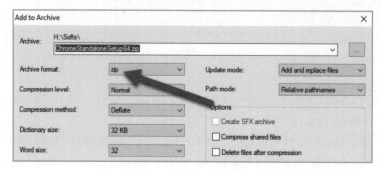

Figure 8-6 "Add to Archive" Window

2. Choose files you wish to extract and press right-click. This will open a new menu.

3. Hover over the "7-Zip" option (Figure 8-7).

Figure 8-7 Hover over the "7-Zip" Option

4. Choose **"Extract Here"**. This brings up a new window that shows the progress as well as the remaining time before extracting the file (Figure 8-8).

5. Wait for it to finish. The extracted file will appear in the same directory where you have all RAR or 7-Zip files in.

Unit 8　Delivering the System　系统交付

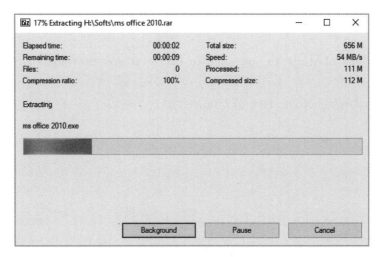

Figure 8-8　Extracting Window

Product Specifications

7-Zip is offering multiple features which makes it one of the best and most effective software compression tools that deliver the lowest compressed data. Few of its remarkable specs are:

- **Open Architecture**—the source code is using GNU LGPL license while the unRAR code is a combination of license with unRAR + GNU LGPL restrictions.
- **High Compression Ratio**—by taking advantage of its own 7-Zip format w/ LZMA as well as LZMA2 compression, it allows 7-Zip to achieve a 40 percent compression ratio or higher compared to its peers. For GZIP and ZIP formats, 7-Zip delivers a compression ratio that's 2 to 10 percent better compared to WinZip and PKZip.
- **Strong AES-256 encryption**—this one includes both encryption and password protection of filenames and files.
- **Ability to use multiple encryptions, conversion or compression methods**—to give you an example, 7-Zip can support several packing as well as unpacking for XZ, GZIP, ZIP, TAR, WIM, and 7-Zip. It also has support for a self-extracting ability for 7-Zip format.
- **Integrates with Windows Shell**—by using the menu options, it is feasible to integrate 7-Zip to the Windows Shell menu.

Conclusion

These are the spec sheet that made 7-Zip a strong competitor among the leading compression tools on the Web. The best part, 7-Zip is available and downloadable free of charge! If you're afraid that you don't know how to use 7-Zip, which stops you from switching, follow the steps above. Start to enjoy the lowest compression at no price.

Exercises

1. What kinds of steps do you need to follow if you download 7-Zip for Mac or Linux?

2. Is 7-Zip downloadable free of charge?

Unit 9

Taking an Interview

参加面试

Listening & Speaking

Unit 9

Dialogue: Interview

(Kevin's application for a position as a programmer in the technical department of Expansion has been responded by the Human Resources Department of the company. Today he comes to the company for an interview.)

Mr. Smith: Good morning, thank you for applying for our opening positions. My name is Smith, the manager of Human Resources Department. What's your name, please?

Kevin: Good morning, Mr. Smith. My name is Kevin Li, and I am coming for the position as a programmer in the technical department.

Mr. Smith: Well, now I would like to talk with you regarding your qualifications for this interview. What can you tell me about yourself? [1]

Kevin: I'm an active and responsible person, and will not give up even if I encounter numerous obstacles on the way to my goal. I like to work in a team, because I believe that it is important to share the opinions with each other to get the best results. I am very interested in the latest advancements in science and technology, and have a good theoretical background and **hands-on** capabilities. I love the IT industry and want to **pursue** an IT career.

Mr. Smith: Can you talk about your education background? [2]

Kevin: I have just graduated from School of Software, BeiHang University this July, and obtained a bachelor degree in software engineering.

[1] Replace with:
1. Tell me something about yourself.
2. How do you **value/judge** yourself?
3. What is your work style?
4. Why do you think that you **deserve** to get the job?

[2] Replace with: What is your education?

Unit 9　Taking an Interview 参加面试

Mr. Smith:	Why do you apply for a job at our company? [3]	[3] Replace with: 1. Why do you want to work for us? 2. What do you know about us?
Kevin:	As we know, Expansion company is one of the longest established and leading software **outsourcing** products, services and solutions providers in China, but also has a global **presence**. So I believe that working here can help me grow professionally. I also think I can contribute to the development of the company.	
Mr. Smith:	Have you ever done an **internship** that helped to prepare you for this type of work? [4]	
Kevin:	Yes. I have participated in several practical projects as a programmer in the Lab which is built by our school's cooperating with Microsoft during my undergraduate study. I participated in a hotel management information system for Four Seasons Hotel as a project manager and a management information system for the financial department of Giant company. Besides, regarding my internship, I have done a research and analysis about the global software market and completed a business report for a domestic leading IT company in English. All of the above are very valuable internship experience for me to prepare for this work, I think.	[4] Replace with: 1. How have you prepared yourself for the transition from school to the workplace? 2. Did you **get your hands on** experience in school? 3. What aspects of your abilities and experience make you think that you will succeed in this job? 4. What attracts you in this job?
Mr. Smith:	Ok, that sounds great! What else do you want to know about your work?	
Kevin:	I want to know what other qualifications you consider necessary for this position?	
Mr. Smith:	Well, as I know, sometimes the **workload** of developing software may be really very heavy, so the person who is in charge of it may have to be under much pressure and must be **energetic** and capable.	

255

Kevin: Yes, I agree with you. On this point, my internship experience has made me used to this kind of work environment.

Mr. Smith: Ok, thank you for your interest in our company. If we decide to bring you onto our team, you will receive an E-mail from us within five working days.

Kevin: Thank you very much. I'm looking forward to hearing from you.

 Exercises

Work in a group, and make up a similar conversation by replacing the statements with other expressions on the right side.

 Words

value['vælju:] v. 评价 judge[dʒʌdʒ] v. 判断,评价 deserve[di'zɜ:v] v. 值得,应得 hands-on 亲身实践的,实际动手做的 pursue[pə'sju:] v. 从事,进行 outsource['autsɔ:s] v. 外包	presence['prezns] n. 势力 internship['intɜ:nʃip] n. 实习生职位,实习期 workload['wɜ:kləud] n. 工作量 energetic[ˌenə'dʒetik] adj. 精力充沛的

 Phrases

get one's hands on   得到,占有

Listening Comprehension: Expansion Company

Listen to the article and answer the following 3 questions based on it. After you hear a question, there will be a break of 15 seconds. During the break, you will decide which one is the best answer among the four choices marked (A), (B), (C) and (D).

Questions

1. What are the core products and/or services offered by Expansion now?
   (A) Testing Service & Software Development
   (B) Software Development & Application Maintenance Service

(C) Software Development & BPO Service

(D) Application Maintenance & BPO Service

2. Who is the largest client of Expansion among its overseas market currently?

(A) Japan

(B) Korea

(C) Europe

(D) U.S.

3. Which point is not mentioned about the future actions of Expansion according to the article?

(A) Increase earning and profit

(B) Develop strong management team

(C) Economize physical resources

(D) Expand new markets

Words

headquarter[ˌhedˈkwɔːtə] v. 总部设于……, 总公司设于……	pool[puːl] n. 资源的集合, 联营
	margin[ˈmɑːdʒin] n. 利润
rich[ritʃ] adj. 宝贵的	advance[ədˈvɑːns] v. 提升, 提高

Abbreviations

BPO Business Process Outsourcing	业务流程外包

Dictation: Software Configuration Management

This article will be played three times. Listen carefully, and fill in the numbered spaces with the appropriate words you have heard.

When you build computer software, ___1___ happens. And because it happens, you need to control it ___2___. Software Configuration Management, SCM for short, is a set of activities that are designed to ___3___ change by identifying the work products that are ___4___ to change, ___5___ relationships among them, defining mechanisms for managing different ___6___ of these work products, controlling changes that are *imposed*, and auditing and ___7___ on the changes that are made. Another name for it is "change control."

257

In his book Code Complete, Steve McConnell **depicts** that: "If you don't control changes to requirements, you can **end up** writing code for parts of the system that are ____8____ eliminated. You can write code that's incompatible with new parts of the system. You might not ____9____ many of the incompatibilities until ____10____ time, which will become **finger-pointing** time because nobody will really know what's going on."

Since programming projects are so **volatile** today, SCM is especially useful to programmers. It focuses on a program's ____11____, source code, ____12____, and test data.

SCM systems are based on a simple ____13____: the **definitive** copies of your files are ____14____ in a central repository. People **check out** copies of files from the ____15____, work on those ____16____, and then check them ____17____ in when they are finished. SCM systems manage and ____18____ **revisions** by multiple people against a single **master** set. All SCM systems provide the following essential ____19____: **concurrency** management, ____20____ and **synchronization**.

Words

impose[imˈpəuz] v. 强制实行,把……强加于
depict[diˈpikt] v. 描述,描写
finger-pointing 指责,责难
volatile[ˈvɔlətail] adj. 不稳定的,易变的
definitive[diˈfinətiv] adj. 最后的,决定性的

revision[riˈviʒn] n. 修改,修订
master[ˈmɑːstə(r)] adj. 支配的,总的,最重要的
concurrency[kənˈkʌrənsi] n. 并发性,并行性
synchronization [ˌsiŋkrənaiˈzeiʃn] n. 同步

Phrases

end up 结束
check out 给……办理结账手续,核实

Abbreviations

SCM   Software Configuration Management   软件配置管理

Unit 9 Taking an Interview 参加面试

Part 2

Reading & Translating

Section A: Building Software Products vs Platforms

In the technology world, many vendors describe their products and services as not just products, but extensible platforms for things like your personal information, for applications, for your smart home, your customer data, and just about anything involving digital information and data. But what really constitutes a platform versus just a product or a system? There are important distinctions, and they are not just about the technology.

Products can be made to be extensible, via published APIs or plug-in architectures, but that does not necessarily make them a platform. Part of making a true platform revolves around the business model to recruit partners, applications, and innovations built around the product, and how value is created around the platform (Figure 9-1).

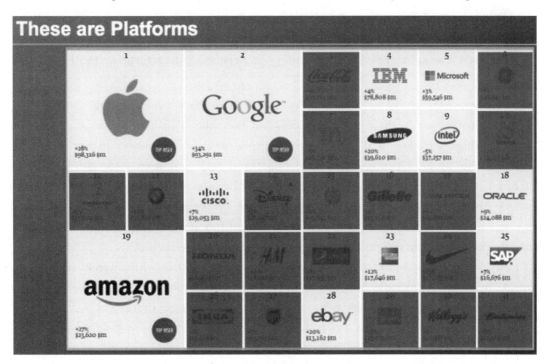

Figure 9-1 These are Platforms

In Three Elements of a Successful Platform Strategy, the authors break down key elements of successful platforms—they call them Connection, Gravity, and Flow.

Connection is about how easily all participants can plug into the platform. Gravity is about how well the platform attracts participants, and Flow is about how the platform allows for creation and exchange of value.

While successful business platforms **predate** the Internet era, today's business platforms are being built upon the rapid growth of cloud, mobile, and social technology. Of course, these technologies (really collections of technologies) have spawned platforms that are being leveraged to build other businesses. The major ones are companies everyone is familiar with—Amazon, Microsoft and Google in cloud services, Apple and Google in mobile, Facebook and Twitter in social.

The Toolbox is typically the technology that creates the connections for participants to plug in to the platform. This can take place at many levels, and is not just about building APIs into a product or service. The APIs will usually be the **building block**, but taking a product to a platform level may require building some key applications that use them.

Operating systems like iOS and Android provide **a host of** APIs for developers but Google and Apple also build some of the most often used apps (Music, E-mail, etc.) on those platforms to **engage** and deepen their users' interactions in those ecosystems.

Yahoo's Flickr photo service has a free app to upload pictures from a PC or mobile device into your Flickr account.

Google's YouTube app allows uploading of personal videos to grow the network of content.

Evernote, Dropbox, and Box APIs allow third party applications to store and retrieve information in their cloud services, but they also build their own.

The Magnet is the "pull" that attracts producers and consumers to the platform. This will usually be a combination of product and technical innovation along with an effective business model. For example, early examples of successful Web platforms include eBay and Skype. eBay offered an online marketplace at a low cost to match buyers and sellers that otherwise would have a hard time finding each other. Part of the model was the review system that let buyers and sellers **rate** each other.

Skype offered a free or low cost way of communicating with anyone around the world, bypassing regulatory long distance fees and other limitations by using the power of the Internet. It grew quickly by **word of mouth** because there was a clear **value proposition** for all participants. Pricing models, reputation systems (rating producers and consumers) and **incentives** for participating in the platform (think **free ride offers** for new users on Uber and Lyft) are all key elements that help **scale** platforms-the gravity that the authors describe.

The Matchmaker is the piece that *fosters* value creation for the participants. Apple's app store created a market for developers to have a ready distribution channel. The store created value for both consumers and app producers—a *curated* shop where consumers can safely get apps, and a ready market for developers to get their apps into the ecosystem. There are many other examples of this today; Google's and Facebook's hugely profitable business models are predicated on making a vast amount of data on user behavior available to advertisers for targeted marketing that previously wasn't available on other platforms like TV, radio, and *print*.

Most of this thinking around platforms today revolves around the three major technology trends in cloud, mobile, and social. However, successful digital product platforms exist on many other levels of scale. Think about Adobe's Photoshop and its plug-in model that has existed for years. The plug-in architecture has helped *cement* its position as the *premier* graphic editing tool for photographers and designers. Microsoft's Office became a platform even before the Internet era, offering developers opportunities to create information workflows and custom applications with Visual Basic, tying together data in spreadsheets, databases, and documents.

More recently, Salesforce.com took Customer Relationship Management (CRM) systems to a platform level by successfully *employing* all these elements—putting the functionality in the cloud, integrating with business and personal social networks, opening up the product to integrations, actively encouraging developers and partnerships to build on top of its platform with a variety of channel programs, and leveraging mobile.

If you are thinking about a platform approach to the problem/need you are trying to address, think about it holistically. The cloud can let you deploy and scale globally, mobile can enable users to access your service or app anywhere, and social can be an effective marketing tool and way to quickly gain adoption.

REST APIs or other hooks [1] into your cloud or app may make it easier for third parties to integrate with your product. But creating a platform means thinking about the value proposition for all participants. It is the combination of technical innovation, business model, and a real value proposition for all participants that creates a platform bigger than the sum of the parts.

Why are Uber and Lyft considered innovative? *In and of itself*, the technology is not that complicated, although as in all things execution is everything. What makes them platforms, *in part*, is that they bring together a large supply of something not easily accessible before (rides in private cars) to a large base of users (through the mobile apps) with a relatively friction-less payment system (credit cards, built into the app). Their technology platforms are cloud based, and they can also allow for different payment systems to integrate as well as other apps (like messaging) to integrate their services.

Words

versus [ˈvɜːsəs] prep. （比较两种不同想法、选择等）与……相对，与……相比
plug-in 插件，外挂程序
while [waɪl] conj. 虽然，尽管
predate [ˌpriːˈdeɪt] v. （在日期上）早于，先于
engage [ɪnˈɡeɪdʒ] v. 吸引住，引起注意
rate [reɪt] v. 评估，评价
incentive [ɪnˈsentɪv] n. 激励，刺激

scale [skeɪl] v. 按比例决定，测量
foster [ˈfɒstə(r)] v. 促进
curate [ˈkjʊərət] v. 操持（收藏品或展品的）展出，组织（音乐节的）演出
print [prɪnt] n. 印刷行业，出版界
cement [sɪˈment] v. 加强，巩固（关系等）
premier [ˈpremɪə(r)] adj. 首要的，最著名的
employ [ɪmˈplɔɪ] v. 使用，应用

Phrases

revolve around　围绕着
building block　组成部分；构成要素
a host of　许多，一大群
word of mouth　口碑，口头传述的
value proposition　价值主张，价值定位
free ride offer　免费乘车优惠
in and of itself　就其本身而言
in part　在某种程度上，部分地

Notes

[1] 钩子（hook）实际上是一个处理消息的程序段，通过系统调用，把它挂入系统。每当特定的消息发出，在没有到达目的窗口前，钩子程序就先捕获该消息，亦即钩子程序先得到控制权。这时钩子程序既可以加工处理（改变）该消息，也可以不作处理而继续传递该消息，还可以强制结束消息的传递。

Exercises

Ⅰ. Read the following statements carefully，and decide whether they are true（T）or false（F）according to the text.

　　____ 1. Uber's technology platform is cloud based.
　　____ 2. Lyft's technology platform is not cloud based.
　　____ 3. Salesforce.com's Office became a platform even before the Internet era.
　　____ 4. Microsoft's app store created a market for developers to have a ready

distribution channel.

_____ 5. Today's business platforms are being built upon the rapid growth of cloud, mobile, and social technology although successful business platforms predate the Internet era.

Ⅱ. Choose the best answer to each of the following questions according to the text.

1. Which of the following technologies is not mentioned in this article?
   (A) social
   (B) O2O
   (C) mobile
   (D) cloud

2. Which of the following is right about Uber and Lyft?
   (A) They can allow for different payment systems to integrate their services.
   (B) Their technology platforms are cloud based.
   (C) They can allow for other apps (like messaging) to integrate their services.
   (D) All of the above.

3. Which of the following is not key element of successful platforms broken down in Three Elements of a Successful Platform Strategy?
   (A) Integration
   (B) Gravity
   (C) Connection
   (D) Flow

Ⅲ. Fill in the numbered spaces with the words or phrases chosen from the box. Change the forms where necessary.

> begin   combine   request   accept   intend
> which   deploy   design   knowledge   release

Release Management

Release management is a software engineering process ___1___ to oversee the development, testing, deployment and support of software ___2___. The practice of release management ___3___ the general business emphasis of traditional project management with a detailed technical ___4___ of the Systems Development Life Cycle (SDLC) and IT Infrastructure Library (ITIL) practices.

Release management usually ___5___ in the development cycle with requests for changes or new features. If the ___6___ is approved, the new release is planned and ___7___. The new design enters the testing or quality assurance phase, in ___8___ the release is built, reviewed, tested and tweaked until it is ultimately ___9___ as a release candidate. The release then enters the deployment phase, where it is implemented and made available. Once ___10___, the release enters a support phase, where bug reports and other issues are collected; this leads to new requests for changes, and the cycle starts all over again.

Ⅳ. Translate the following passage into Chinese.
Software Architecture

Software architecture is a structured framework used to conceptualize software elements, relationships and properties. This term also references software architecture documentation, which facilitates stakeholder communication while documenting early and high-level decisions regarding design and design component and pattern reuse for different projects. The software architecture process works through the abstraction and separation of these concerns to reduce complexity.

Section B: Secure Software Development Life Cycle

When it comes to creating, releasing, and maintaining functional software, most organizations have a well-oiled machine in place.

However, when it comes to securing that software, not so much. Many development teams still perceive security as interference—something that throws up hurdles and forces them to do rework, keeping them from getting cool new features to market.

But insecure software puts businesses at increasing risk. Cool new features aren't going to protect you or your customers if your product offers exploitable vulnerabilities to hackers. Instead, your team needs to integrate security into the entire Software Development Life Cycle (SDLC) so that it enables, rather than inhibits, the delivery of high-quality, highly secure products to the market.

A SDLC is a framework for the process of building an application from inception to decommission (Figure 9-2). Over the years, multiple SDLC models have emerged—from waterfall and iterative to, more recently, agile and CI/CD, which increase the speed and frequency of deployment.

In general, SDLCs include the following phases:
- Planning and requirements
- Architecture and design
- Test planning
- Coding
- Testing and results
- Release and maintenance

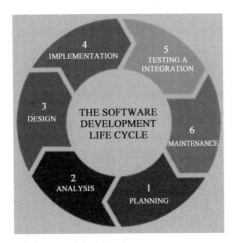

Figure 9-2　A SDLC

In the past, organizations usually performed security-related activities only as part of testing—at the end of the SDLC. As a result of this **late-in-the-game** technique, they wouldn't find bugs, flaws, and other vulnerabilities until they were far more expensive and time-consuming to fix. Worse yet, they wouldn't find any security vulnerabilities at all.

The Systems Sciences Institute at IBM reported that it cost six times more to fix a bug found during implementation than one identified during design. Furthermore, according to IBM, the cost to fix bugs found during the testing phase could be 15 times more than the cost of fixing those found during design.

So it's far better, not to mention faster and cheaper, to integrate security testing across the SDLC, not just at the end, to help discover and reduce vulnerabilities early, effectively **building** security **in**. Security assurance activities include architecture analysis during design, code review during coding and build, and **penetration testing** before release. Here are some of the primary advantages of a secure SDLC approach:

- Your software is more secure, as security is a continuous concern.
- All stakeholders are aware of security considerations.
- You detect design flaws early, before they're coded into existence.
- You reduce your costs, thanks to early detection and resolution of defects.
- You reduce overall **intrinsic** business risks for your organization.
- How does a secure SDLC work?

Generally speaking, a secure SDLC involves integrating security testing and other **activities** into an existing development process. Examples include writing security requirements **alongside** functional requirements and performing an architecture risk analysis during the design phase of the SDLC.

Many secure SDLC models are in use, but one of the best known is the Microsoft Security Development Lifecycle (MS SDL), which outlines 12 practices organizations can adopt to increase the security of their software. And before, NIST [1] published the final version of its Secure Software Development Framework, which focuses on security-related processes that organizations can integrate into their existing SDLC.

If you're a developer or tester, here are some things you can do to move toward a secure SDLC and improve the security of your organization:

- Educate yourself and co-workers on the best secure coding practices and available frameworks for security.
- Conduct an architecture risk analysis at the start.
- Consider security when planning and building for test cases.
- Use code scanning tools for static analysis, dynamic analysis, and interactive application security testing.

Beyond those basics, management must develop a strategic approach for a more significant impact. If you're a decision-maker interested in implementing a complete secure SDLC **from scratch**, here's how to get started:

- Perform a **gap analysis** to determine what activities and policies exist in your organization and how effective they are.
- Create a Software Security Initiative (SSI) by establishing realistic and achievable goals with defined metrics for success.
- Formalize processes for security activities within your SSI.
- Invest in secure coding training for developers as well as appropriate tools.
- Use outside help as needed.

Does your organization already follow a secure SDLC? Fantastic, well done! But there's always **room** for improvement. One way to determine your standing is by evaluating your security program against real-life programs at other organizations. The Building Security In Maturity Model (BSIMM) can help you do that. For the past decade, the BSIMM has tracked the security activities performed by more than 100 organizations. Because every organization and SDLC is different, the BSIMM doesn't tell you exactly what you should do. But its observational model shows you what others in your own **industry vertical** are doing—what's working and what isn't.

Words

well-oiled（系统或机构）运转顺畅的，运行良好的
cool[ku:l] *adj.* 棒的，酷的
exploitable[ɪksˈplɔɪtəbl] *adj.* 可利用的，有利可图的
inhibit[ɪnˈhɪbɪt] *v.* 阻止，阻碍
decommission[ˌdiːkəˈmɪʃn] *v.* 停止使用，使退役

late-in-the-game 比赛后期
intrinsic[ɪnˈtrɪnzɪk] *adj.* 固有的，内在的，本身的
alongside[əlɒŋˈsaɪd] *prep.* 与……一起，与……同时
beyond[bɪˈjɒnd] *prep.* 除……之外
room[ru:m] *n.* 余地，空间

Unit 9 Taking an Interview 参加面试

 Phrases

come to　谈到，涉及
throw up　突然建造，匆忙建造
build in　使成为一部分，嵌入
penetration testing　渗透测试，渗透检查
generally speaking　一般说来
from scratch　从头做起，白手起家
gap analysis　差异分析
industry vertical　产业垂直

 Notes

[1]　美国国家标准与技术研究院（National Institute of Standards and Technology，NIST）直属美国商务部，从事物理、生物和工程方面的基础和应用研究，以及测量技术和测试方法方面的研究，提供标准、标准参考数据及有关服务，在国际上享有很高的声誉。

 Exercises

Ⅰ. Read the following statements carefully，and decide whether they are true（T）or false（F）according to the text.

　　____ 1. If you're a developer or tester，you can consider security when planning and building for test cases.
　　____ 2. If you're a decision-maker interested in implementing a complete secure SDLC from scratch，you can use outside help as needed.
　　____ 3. According to IBM，the cost to fix bugs found during the testing phase could be 100 times more than the cost of fixing those found during design.
　　____ 4. According to IBM，the cost to fix bugs found during implementation could be 12 times than one identified during design.
　　____ 5. MS SDL outlines 12 practices organizations can adopt to increase the security of their software.

Ⅱ. Choose the best answer to each of the following questions according to the text.

　1. How many practices does MS SDL outline organizations can adopt to increase the security of their software?
　　（A）Six
　　（B）Eight
　　（C）Ten

(D) Twelve

2. What is the abbreviation of BSIMM?
   (A) Building Security In Material Model
   (B) Building Software In Maturity Model
   (C) Building Software In Material Model
   (D) Building Security In Maturity Model

3. Which of the following is not included in SDLCs?
   (A) Marketing
   (B) Release and maintenance
   (C) Coding
   (D) Testing and results

Ⅲ. Fill in the numbered spaces with the words or phrases chosen from the box. Change the forms where necessary.

> in　like　to　many　collect
> use　as　report　name　help

Software Measurement in Software Engineering

To assess the quality of the engineered product or system and to better understand the models that are created, some measures are ___1___. These measures are ___2___ throughout the software development life cycle with an intention to improve the software process on a continuous basis. Measurement ___3___ in estimation, quality control, productivity assessment and project control throughout a software project. Also, measurement is used by software engineers to gain insight into the design and development of the work products. ___4___ addition, measurement assists in strategic decision-making as a project proceeds.

Software measurements are of two categories, ___5___, direct measures and indirect measures. Direct measures include software processes like cost and effort applied and products ___6___ lines of code produced, execution speed, and other defects that have been ___7___. Indirect measures include products like functionality, quality, complexity, reliability, maintainability, and ___8___ more.

Generally, software measurement is considered ___9___ a management tool which if conducted in an effective manner, helps the project manager and the entire software team to take decisions that lead ___10___ successful completion of the project.

Ⅳ. **Translate the following passages into Chinese.**

Software Quality Assurance (SQA)

SQA helps ensure the development of high-quality software. SQA practices are implemented in most types of software development, regardless of the underlying software development model being used. In a broader sense, SQA incorporates and implements software testing methodologies to test software. Rather than checking for quality after completion, SQA processes test for quality in each phase of development until the software is complete. With SQA, the software development process moves into the next phase only once the current/previous phase complies with the required quality standards.

SQA generally works on one or more industry standards that help in building software quality guidelines and implementation strategies. These standards include the ISO 9000 and Capability Maturity Model Integration (CMMI).

Part 3

Simulated Writing: Resume

A resume is the key job search tool and should be limited to one page—or two pages if you have substantial experience. Because the resume gives potential employer's a first impression of you, take the time to make sure that it is well organized, well designed, easy to read, and free of errors. Only use high-grade top quality paper.

Analyzing Your Background

Determine what kind of job you are seeking and what information about you and your background would be most important and related to that kind of job. List the following:

- Schools attended, degrees, major field of study, academic honors, grade point average (if recently graduated), particular and relevant academic projects.
- Jobs held, primary and secondary duties in each of them, when and how long you held each job, promotions, skills you developed in your jobs.
- Other experiences and skills you have developed that would be of value in the kind of job you are seeking; extracurricular activities that have contributed to your learning experience; leadership, interpersonal, and communication skills you have developed; any collaborative work you have performed; computer skills you have acquired.

Organizing Your Resume

Most resumes have information arranged chronologically in the following

categories:
- Heading (name and contact information)
- Job Objective (optional)
- Education
- Employment Experience
- Computer Skills (optional)
- Honors and Activities
- References

If you are recent graduate without much work experience, list education first. If you have many years of job experience, including jobs directly related to the kind of position you are seeking, list employment experience first. In your education and employment sections, use a chronological sequence with most recent experience first, the next most recent experience second, and so on.

- **Heading**

At the top of your resume, include all of your contact information including: name, address, telephone, fax, and E-mail address. Make sure that your name stands out on the page. If you have just one address, center it at the top of the page. If you have two addresses such as school and home or temporary and permanent, place school or temporary address on the left side of the page and your permanent or home address on the right side. Place both addresses underneath your name.

- **Objective**

A job objective introduces the material in a resume and helps the reader quickly understand your goal. Write your objective in three lines or less. For example: a full-time software programmer/analyst aimed at solving engineering problems.

- **Education**

List the school(s) you have attended, the degrees you received and the dates you received them, your major field(s) of study, and any academic honor you have earned. Include your grade point average only if it is 3.0 or higher. List courses only if they are unusually impressive or if your resume is otherwise sparse.

- **Employment Experience**

You can organize your employment experience chronologically, starting with your most recent job and working backward under a single major heading called "Experience", or you could also organize your experience functionally by clustering similar types of jobs into one or several sections with specific headings such as "Management Experience" or "Major Accomplishments".

- Include jobs or internships when they relate directly to the position you are seeking.
- Include extracurricular experiences, especially those involving a leadership position or community-service.

- Under each job or experience, provide a concise description of your primary and secondary responsibilities.
- Use action verbs with precision and conciseness. Do not use "I".
- **Computer Skills**

If you are applying for a job that requires computer knowledge, include specific languages, software, and hardware you are familiar with.

- **Honors and Activities**

If there is still room on the resume (less than 2 pages), you can add items such as fluency in foreign languages, writing and editing abilities, specialized technical knowledge, student or community activities, professional or club memberships, and published works. Be selective: do not duplicate information given in other categories and include only information that supports your employment objective.

- **References**

Unless your resume is sparse, avoid listing references. For an interview, have a printed list of references available and bring them with you. Your list should include the main heading "References for [your name]".

P. S. This guide can be tailored according to the different resumes.

A Sample Resume

Kevin Li

School of Software

Beihang University，Beijing 100191，P. R. China

kevin_li@buaa.edu.com

0086-13911234567

Objective

To obtain a challenging position as a software engineer with an emphasis in software design and development.

Education

Bachelor of Engineering	7/2021

School of Software

Beihang University，Beijing

Major：Software Engineering

GPA：3.84/4.00

CAREER RELATED EXPERIENCE	**System Designer & Programmer**	1/2021 - 6/2021

Digital Earth & GIS Lab，Beihang University，Beijing

- Designed and developed UI of a platform of Digital Earth sponsored by National Fund
- Designed and developed a 3D simulation system for airplane

System Designer & Programmer	9/2020 - 1/2021

School of Software，Beihang University，Beijing

- Designed and developed Resource Management System for Financial Office of School
- Designed and developed Information Management System for Library of School

ADDITIONAL EXPERIENCE	**Market Research Analyst（Intern）**	7/2020 - 8/2020

Giant Software，Inc.，Beijing

- Did research and analysis on global and domestic markets of outsourcing software in a complete English environment
- Wrote a business report and plan for the company in English

Teaching Assistant for C++ Programming	3/2020 - 6/2020

School of Software，Beihang University，Beijing

- Assisted teacher to judge the assignments for students
- Instructed students to practice C++ programming on PCs

SKILLS	Language: CET-6, fluency in written and spoken English
	Computer: Having rich and systematic knowledge in computer and software engineering domain
	Be familiar with Python, C++, and Java programming
	Mastering a variety of UI design software
HONORS	First Award of "Student Scholarship", Beihang University, 2020
	Third Award of "Creative Technology", Beihang University, 2020
REFERENCES	Available upon request

 Exercises

There are two scenarios below, please choose one and write a resume.

Scenario 1: As an undergraduate in grade 4, you want to apply an internship position in the R & D center, a financial company.

Scenario 2: With 2 years event planning experience, you want to join in the University Student Union.

Unit 10

Beginning Your Work

开始工作

Unit 10　Beginning Your Work 开始工作

Part 1

Listening & Speaking

Unit 10

Dialogue：Beginning Your Work

（*Congratulations on Kevin becoming a new employee as a programmer in the Information System Department of Expansion Company！Today is his first day for work．There will be some new colleagues and fresh working environment coming along with him．*）

Kevin： Nice to meet you，Mr. John. My name is Kevin Li，and I'd like to report for work.

Mr. John： (the CTO of Expansion)：Welcome to our company，Kevin. You'll work in the Information System Department. Ok，now please let me introduce our colleagues to you.

（*Kevin comes into the office following Mr．John．*）

Mr. John： Hello，everybody. Today I'd like to introduce a new colleague to you. His name is Kevin Li and he will work with you from today.

Kevin： How do you do?

A，B，C： How do you do? Mr. Li，welcome！

Kevin： Thanks! I have just graduated. This is my first time working in a multinational company. I hope you can help me adapt from school life to working in a professional environment.

A： No problem，Mr. Li. I'm Jenny，in charge of[1] the office work. This is your desk.

Kevin： Thank you very much，Ms. Jenny.

[1] Replace with：
1. take charge of
2. be responsible for

275

A: You call me Jenny. Everybody in our company is called by his or her first name. It's been our tradition since the company was established.

Kevin: Ok, I'll try.

B & C: Very glad to know you. We believe we will be able to cooperate well in the future.

Kevin: So do I.

Mr. John: Kevin, if you have any questions please feel free. The people here are all very kind. Any other questions?

Kevin: All right. By the way, excuse me. Would you please tell me what my specific tasks and responsibilities are? [2]

[2] Replaced with: Do you mind if I ask you something about my work responsibilities in the company?

Mr. John: You'll spend most of this week to get familiar with the work in the office and read some documents about a project which you will participate in from the next week. Do you have any other questions?

Kevin: Not yet, but I'm sure I will have more questions later. Thanks.

(*It is time for ringing out after have been working for a whole day.*)

Jenny: Kevin, you've been working for nearly one day. How do you feel about the job? [3]

[3] Replace with:
1. How are you feeling about the job?
2. What do you think about the job?

Kevin: Not bad. Thank you for your help. I've been busy all day and feeling a bit tired.

Jenny: I had the same feeling when I first came to work here. But after a period of time, I felt much better. I'm sure you'll get used to the job.

Unit 10　Beginning Your Work　开始工作

Kevin: I also feel that the work efficiency is very high and it is also challenging. Everybody here works hard and they are cooperative. And you all have strong capabilities and professional skills. I really like such a working environment.

 Exercises

Work in a group, and make up a similar conversation by replacing the statements with other expressions on the right side.

 Phrases

adapt to　适应
not yet　还没有
ring out　（根据考勤钟）打卡下班
get used to　习惯

 Abbreviations

CTO　Chief Technology Officer　首席技术官

Listening Comprehension: The Organizational Structure of a Company

Listen to the article and answer the following 3 questions based on it. After you hear a question, there will be a break of 15 seconds. During the break, you will decide which one is the best answer among the four choices marked (A), (B), (C) and (D).

Questions

1. How many major leading positions in most companies are mentioned in the article?
   (A) Four　　　　　　　　(B) Five
   (C) Six　　　　　　　　 (D) Seven

2. Whom should you be directly under the supervision of if you are a software engineer in an IT company?
   (A) CEO　　　　　　　　(B) CTO
   (C) COO　　　　　　　　(D) CSO

3. Which task does not likely belong to the responsibilities of the CFO according to the article?
   (A) Recruiting a staff member
   (B) Analyzing the margins changing trend
   (C) Taking revenue forecast
   (D) Making a financial planning

 Words

subordinate[sə'bɔːdinət, sə'bɔːdineit] n. 下级,部属
chief[tʃiːf] n. 头目,首领
operation[ˌɔpə'reiʃn] n. 运营,经营
board[bɔːd] n. 董事会,理事会
oversee[ˌəuvə'siː] v. 管理,监督
acronym['ækrənim] n. 只取首字母的缩写词
class[klɑːs] v. 把……归入某等级,把……看作
rank[ræŋk] n. 等级
besides[bi'saidz] adv. 此外

 Phrases

take the lead   带头,一马当先
for short   作为简称,作为缩写
under the supervision of   在……的监督下,在……的指导下

 Abbreviations

CFO   Chief Financial Officer   首席财务官,财务总监
COO   Chief Operating Officer   首席运营官
CBDMO   Chief Business Development and Marketing Officer   首席商业扩展和营销官
CSO   Chief Strategy Officer   首席策略官

Dictation: Open Source Movement

This article will be played three times. Listen carefully, and fill in the numbered spaces with the appropriate words you have heard.

In the early days of computing, software came ____1____ with the computer, including the source code for the software. Programmers adjusted and adapted the programs and happily shared the improvements they made. In the 1970s, firms began **withholding** the ____2____ code, and software became big ____3____.

Unit 10 Beginning Your Work 开始工作

 As opposed to being kept ___4___, "open source" means that code is freely ___5___ by those who write it, which ___6___ the power of distributed peer review and transparency of process. The promise of open source is better quality, higher ___7___, more flexibility, lower cost, and an end to **predatory** vendor lock-in[1].

 In 1960's, researchers with ___8___ to the Advanced Research Projects Agency Network (**ARPANET**) used a process called Request for Comments to develop telecommunication network **protocols**. ___9___ by contemporary open source work, this collaborative process **led to** the ___10___ of the Internet in 1969. The decision in the free software movement to use the label "open source" **came out of** a strategy **session** ___11___ at Palo Alto, California, **in reaction to** Netscape's January 1998 ___12___ of a source code ___13___ for Navigator.

 So far, much of the ___14___ programming that **powers** the Internet, creates ___15___, and produces software is the result of "open source" code. Leaving source code open has ___16___ some of the most sophisticated developments in computer technology, including, most **notably**, Linux and Apache, which **pose** a significant ___17___ to Microsoft in the marketplace.

 In his book The Success of Open Source, Steven Weber argues that the success of open source is not a **freakish** ___18___ to economic principles. The open source community is guided by standards, rules, decision-making ___19___, and **sanctioning** ___20___.

◆ Words

withhold[wɪð'həʊld] v. 拒绝给，不给	power['paʊə(r)] v. 促进，推动，驱动
predatory['predətri] adj. 掠夺性的，欺负弱小的，压榨他人的	notably['nəʊtəbli] adv. 尤其，特别，极大程度上
protocol['prəʊtəkɒl] n. (数据传递的) 协议，规程	pose[pəʊz] v. 引起，造成
	freakish['friːkɪʃ] adj. 反常的，奇怪的
session['seʃn] n. (议会等的) 会议	sanction['sæŋkʃn] v. 批准，认可，同意

◆ Phrases

as opposed to	与……截然相反，对照
lead to	导致
come out of	出自
in reaction to	应对

279

Abbreviations

ARPANET　Advanced Research Projects Agency Network　美国高级研究计划署网络

Notes

[1] 厂商陷阱(Vendor lock-in,供应商套牢,供应商陷阱,厂商泥潭):使一个系统过于依赖于外部所提供的组件/部件,意思是你采用了一个技术,即将自己锁定在这家供应商身上,不能轻易转换供应商。

Part 2

Reading & Translating

Section A: The Organizational Structure of a Company

Consider any **sizable** organization with which you are familiar. Its purpose is to perform a service or deliver a product. If it's nonprofit, for example, it may deliver the service of educating students or the product of food for **famine** victims. If it's profit-oriented, it may, for example, sell the service of fixing computers or the product of computers themselves. Information—whether computer-based or not—has to flow within an organization in a way that will help managers, and the organization, achieve their goals. **To this end**, organizations are often structured horizontally and vertically—horizontally to reflect functions and vertically to reflect management levels.

Depending on the services or products they provide, most organizations have departments that perform six functions: Research and Development (R & D), production (or operations), marketing and sales, accounting and finance, human resources (personnel) and information systems (IS) (Figure 10-1).

Research and development: The Research and Development (R & D) department does two things: (1) It conducts basic research, relating discoveries to the organization's current or new products. (2) It does product development and tests and modifies new products or services created by researchers.

Production (operations): The production department makes the product or provides the service. In a manufacturing company, it takes the raw materials and has people or machinery turn them into finished goods. In another type of company, this department might manage the purchasing, handle the inventories, and control the flow of goods and

Unit 10 Beginning Your Work 开始工作

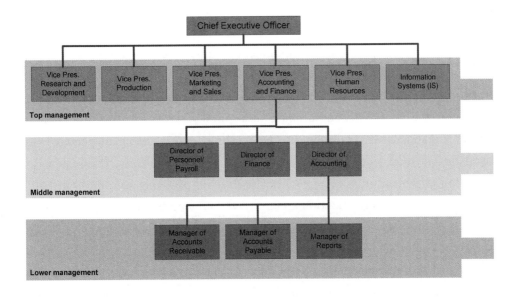

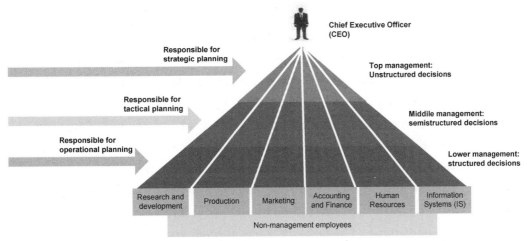

Figure 10-1 Organization Chart

services.

  Marketing and sales: The marketing department oversees advertising, **promotion**, and sales. The people in this department plan, **price**, advertise, promote, **package**, and distribute the services or goods to customers or clients.

  Accounting and finance: The accounting and finance department handles all financial matters. It handles cash management, pays **bills** and taxes, **issues paychecks**, records payments, makes investments, and compiles **financial statements** and reports. It also produces financial budgets and forecasts financial performance after receiving information from other departments.

Human resources: The human resources, or personnel, department finds and hires people and administers sick leave and retirement matters. It is also concerned with compensation levels, professional development, employee relationship, and government regulations.

Information systems (IS): The IS department manages the organization's computer-based systems and plans for and purchases new ones.

Large organizations traditionally have three levels of management—strategic management, tactical management, and operational management. These levels can be shown on an organization chart, a schematic drawing showing the hierarchy of formal relationships among an organization's employees. Managers on each of the three levels have different levels of responsibility and are therefore required to make different kinds of decisions (Figure 10-1).

Strategic-level management: Top managers are concerned with long-range, or strategic, planning and decisions. This top level is headed by the chief executive officer (CEO) along with several vice presidents or managers with such titles as Chief Financial Officer (CFO), Chief Operating Officer (COO), and Chief Information Officer (CIO). Strategic decisions are complex decisions rarely based on predetermined routine procedures; they involve the subjective judgment of the decision maker. For instance, strategic decisions relate to how growth should be financed and what new markets should be tackled first. Determining the company's 5-year goals, evaluating future financial resources, and formulating a response to competitors' actions are also strategic decisions.

Tactical-level management: Tactical, or middle-level managers, make tactical decisions to implement the strategic goals of the organization. A tactical decision is made without a base of clearly defined informational procedures; it may require detailed analysis and computations. Examples of tactical-level managers are plant manager, division manager, sales manager, branch manager, and director of personnel.

Operational-level management: Operational, or low-level (supervisory level), managers make operational decisions—predictable decisions that can be made by following well-defined sets of routine procedures. These managers focus principally on supervising non-management employees, monitoring day-to-day events, and taking corrective action where necessary. An example of an operational-level manager is a warehouse manager in charge of inventory restocking.

Note that, because there are fewer people at the level of top management and many people at the bottom, this management structure resembles a pyramid. It's also a hierarchical structure because most of the power is concentrated at the top.

Unit 10　Beginning Your Work　开始工作

 Words

sizable['saizəbl] *adj.* 相当大的,很大的
famine['fæmin] *n.* 饥荒,奇缺
promotion[prə'məuʃn] *n.* （商品等的）宣传,推销（活动）
price[prais] *v.* 给……定价,（在商品上）标价
package['pækidʒ] *v.* 包装（商品）
bill[bil] *n.* 账单,（餐馆的）账单
issue['iʃu:] *v.* 发出,（正式）发给
paycheck['peitʃek] *n.* 薪金,工资,（支付薪金或工资的）支票
administer[əd'ministə(r)] *v.* 管理,掌管
schematic[ski:'mætik] *adj.* 示意性的,图示的

long-range 长远的,长期的
head[hed] *v.* 领导,主管
subjective[səb'dʒektiv] *adj.* 主观的,个人的
finance['fainæns] *v.* 供给……经费,负担经费
tackle['tækl] *v.* 处理,解决
principally['prinsəpli] *adv.* 主要地,大部分地
restock[ˌri:'stɔk] *v.* 补充（货源）,再补给
resemble[ri'zembl] *v.* 看起来像,显得像
pyramid['pirəmid] *n.* 金字塔形的物体（或一堆东西）,金字塔

 Phrases

to this end　为了这个目的,为了达到这个目标
financial statement　财务报表
sick leave　（尤指拿病假工资的）病假

 Abbreviations

R & D　Research and Development　研究与开发
CIO　Chief Information Officer　首席信息官

 Exercises

Ⅰ. Read the following statements carefully，and decide whether they are true（T）or false（F）according to the text.

　　____1. In the horizontal view，different departments of an organization are partitioned based on different functions.

　　____2. There is a horizontal structure within each department of an organization.

　　____3. Strategic decisions are more long-term than tactical decisions.

　　____4. Tactical decisions don't require detailed analysis and computations.

_____ 5. Well-defined sets of routine procedure can help to make predictable operational decisions.

Ⅱ. **Choose the best answer to each of the following questions according to the text.**

1. Which of the following is wrong about the organizational structure of a company according to this article?
   (A) Information always flows from the top to the bottom within a pyramidal organization.
   (B) The IS department is generally parallel to the marketing and sales department.
   (C) Sales managers belong to the marketing department horizontally and tactical-level of management vertically.
   (D) Non-management employees are supervised directly by managers on operational-level.

2. Which of the following is right about the responsibilities of different departments?
   (A) The R & D department conducts basic research and makes investments.
   (B) The production department relates the discoveries in research to the organization's current or new products. conducts relative discoveries in research to the organization's current or new products.
   (C) The marketing department spreads the services or goods to the public.
   (D) The human resources department issues salaries.

3. Which of the following is wrong about the levels of management in an organization?
   (A) Decisions made on lower levels are to implement the superior goals.
   (B) Fewer people hold most of the power.
   (C) Staffs on each of levels have different levels of responsibility.
   (D) Strategic decisions are made by the completely objective judgment of the superiors on the top level.

Ⅲ. **Fill in the numbered spaces with the words or phrases chosen from the box. Change the forms where necessary.**

> give do far affect discuss
> exist concern require undergo update

Software Reengineering

Before we begin ___1___ about software reengineering, let us get to software engineering first. Software engineering is a discipline of engineering that is ___2___ with the design, development, testing, maintenance, and deployment of a software product. These preceding aspects are part of the Software Development Life Cycle (SDLC) that a software ___3___ before finally made available for clients and users. ___4___ that information, we can now discuss what Software Reengineering is.

Software Reengineering is the process of ___5___ software without ___6___ its functionality. This process may be ___7___ by developing additional features on the software and adding functionalities that may or may not be ___8___ but considered to make the software experience better and more efficient. As ___9___ as the definition goes, this process also entails that the software product will have improved maintainability. Thus, software reengineering is a step towards continuous improvement of software for it to be handled better by developers and clients alike. Additionally, it is a way to make ___10___ products continue in service.

Ⅳ. Translate the following passages into Chinese.

Importance of Good Communication

What makes this interesting is that it's really essential to have good communication methods and skills in software development. Passing information about specifications, requirements, bugs in the code etc. is really crucial for the developer, the product and the business. There must be strong common understanding between the requester and the one actually doing the work. Otherwise you might end up with bad or even completely wrong solution for the problem at hand.

Many times information is lost no matter what communication method is used. To be sure all necessary information is known to people you should ask questions or test them otherwise. This might be difficult especially if the audience is large.

Section B: Why We Need to Address Ethical Issues in Software Engineering

When we think about software development, ethics isn't always the first thing to pop into our minds (Figure 10-2). After all, when creating software, you as a developer are supposed to address technical questions such as functionality and project specifications. But what we usually fail to recognize is that software and technology affect people's lives on a personal level and have the power to make them either better or worse.

Everything people do today involves some kind of software. Driving a car, buying food, communicating, commuting, watching TV, shopping online—the list goes on and on. These technologies power our lives and are inseparable from human life.

Technology has even changed how businesses operate. In order to be the first to introduce

Figure 10-2　Ethical Issues in Software Engineering

products to the market, the best in development, most innovative in products and services, many enterprises overlook the side effects of their ventures and the issues they may cause to people's lives.

Let's face it—there are companies that don't play by the rules. Here is not pointing fingers, but it's a fact that companies hungry for profit do overlook common ethical business standards. Software developers get involved in unethical practices by working for cheating companies that put their own profit above people's lives and turn away from the consequences.

With technology being a huge part of our daily lives, you can't separate it from the ethics that affect daily life. They shape how we consume and how we create.

That being said, it is the responsibility of software engineers to provide users with a secure and transparent program that they can trust. After all, with great power comes great responsibility.

At first sight, it is easy to think that the technical part of development is not directly related to people's lives. After all, it is the business practices that really affect users. However, software developers are the ones who know what their products can do.

As consumers, we trust software providers to help us to optimize and improve our day-to-day lives in exchange for information like our name and E-mail address. As software developers, we trust software providers to make good use of our creations. This trust is supported by vague regulations that still have a lot of loopholes if you know where to look. No one is completely protected from the unfair practices of software providers.

Here are some of the ethical problems that software developers should be aware of when creating their products and choosing which companies to work for.

(1) Protection of Customer Data

Numerous websites' services exist in large part to gather your information.

Take Google for instance. Here is some of the information it has about you: where you've been, your search history, applications you've used and with whom you use them, your YouTube history, etc. Google even allows you to download all the data they have about you—in all honesty, it would fill a huge number of Word documents!

What happens if the government or another legal entity demands information on clients from the information you've gathered with software you built? Where does your moral commitment lie? Have you conveyed your strategies clearly to your clients, and how have you secured their information?

Personal data security is one of the biggest concerns in the digital world because of the sensitive information that your clients trust you with. Personal information is a point of interest for many organizations, from national security to cybercriminals. The companies that don't have policies in place for how to act in these kinds of situations place their clients at risk by not informing them of how their data will be treated.

(2) Intellectual Property

In the fast-growing and profoundly aggressive innovation industry, software engineers and proprietors should practice caution to guarantee that their creations are properly protected inside the system of innovation rights.

Software development deals with interrelated issues that consist of a blend of copyright, patent, trademark, and competitive advantages law. To make sure that the customers are protected from unethical business practices, people in the software development business should be familiar with all these factors and how to implement them effectively. Unethical practices caused by the lack of knowledge do not excuse software developers from taking the responsibility for how their actions affect people's quality of life.

(3) Ownership of Copyright

In theory, you as a software developer own the copyright to your software creation, and no one can copy, distribute, display, or make changes to it without your permission. When it is created by a third-party engineer or a development agency, copyright agreement should always be involved in the collaboration process to define and protect the rights of the code creator and the client who originally had the idea. Registering copyright prevents your business from getting into trouble over ownership rights.

Typically, after the product has been commercially published, the source code is kept confidential to protect it from illegal copying and distribution. Using copyright to protect your source code is beneficial for the company because it provides a convenient way of securing intellectual property rights.

（4）License Agreement

If a client requires a software engineer to convey source code, the parties should clarify whether the client needs to claim the source code or simply modify or update the product later.

If the parties agree to a product permits agreement where the source code is required to be revealed to redo or refresh the product, the engineer may incorporate an arrangement under which the client is committed to keeping the source code confidential.

These aren't the only ethical issues to watch out for. From hackers and cybercriminals to companies overlooking errors, these all fall under concerns about the state of ethics in the software engineering world.

These issues may seem distant and unrealistic until you actually face them. The truth is that no one is 100% protected, which is exactly why ensuring that companies follow software development codes of ethics and avoid shady practices should concern everyone from consumers to developers themselves.

（5）Ethical Solutions

The tricky part about ethical questions is that they address a person's own moral code that has been formed through years of education, family, and societal impact. Add to it that life is not always black and white, and you have yourself a nice brain-twisting puzzle.

Even though facing these ethical dilemmas as software developers seems tricky, there are solutions and steps we can take to do better.

（6）Ethics Education

The most important step to take is to educate yourself and other software engineers about the ethics of your work. Development bootcamps often skip this part in favor of practical knowledge that can be used directly at work.

Technology is not neutral. Educating people about ethical issues and the consequences of their actions has become crucial at this point. The mindset of "Do it now, ask for forgiveness later" cannot be ruling business practices.

How the company operates is decided by management, not software engineers, and you can be forced to follow the chosen course of action even if you don't agree with it. Educating software engineers about the universal standards of business and software development ethics can improve their understanding of their responsibility to society and how to act on it. It can also help you decide which companies to work for or how to behave ethically in your own business.

 **Words**

loophole[ˈluːphəul] n. （法律、合同等的）漏洞,空子	proprietor[prəˈpraiətə(r)] n. 业主,所有人
profound[prəˈfaund] adj. 意义深远的,深刻的,深奥的	distribute[diˈstribjuːt] v. 分销
	claim[kleim] v. 要求（拥有）,索取,索要

Unit 10　Beginning Your Work 开始工作

code[kəud] *n.* 道德准则，行为规范
shady[ˈʃeidi] *adj.* 非法的
tricky[ˈtriki] *adj.* 难办的，难对付的
brain-twisting 费脑的
puzzle[ˈpʌzl] *n.* 不解之谜，疑问
dilemma[diˈlemə] *n.* （进退两难的）窘境，困境

mindset[ˈmaindset] *n.* 观念模式，思维倾向
management[ˈmænidʒmənt] *n.* 经营者，管理部门

 Phrases

pop into one's mind　突然想到
go on and on　一直继续下去，说个不停
side effect　副作用，意外的连带后果
cause to　导致，造成，带来
play by the rules　遵守规则，按规则出牌
pointing fingers　指手画脚
hungry for　渴望
get involved in　涉及，卷入
turn away from　回避
that being said　尽管如此，另一方面
at first sight　乍一看，初看时
in exchange for　作为交换
be protected from　被保护以免受
in large part　大体上，在很大程度上
in all honesty　老实说
trust sb. with sth./sb.　托付，托交，把……委托给某人照管
inform of　将……告知（某人）
watch out for　密切注意，戒备，提防
fall under　归入……类，属于
add to it　加上
development bootcamp　开发训练营
act on　按照……而行动，对……起作用

Exercises

Ⅰ. Read the following statements carefully，and decide whether they are true（T）or false（F）according to the text.

____ 1. Technology is neutral.
____ 2. The mindset of "Do it now，ask for forgiveness later" can be ruling business

practices.

_____ 3. Usually, the source code is kept confidential to protect it from illegal copying and distribution after the product has been commercially published.

_____ 4. Technical part of development is directly related to people's lives.

_____ 5. It is not the responsibility of software engineers to provide users with a secure and transparent program that they can trust.

Ⅱ. Choose the best answer to each of the following questions according to the text.

1. Which of the following is the ethical problem that software developers should be aware of when creating their products and choosing which companies to work for?
   (A) Ownership of copyright
   (B) License agreement
   (C) Protection of customer data
   (D) All of the above

2. Which of the following is wrong about ownership of copyright?
   (A) Registering copyright prevents your business from getting into trouble over ownership rights.
   (B) Personal data security is not one of the biggest concerns in the digital world because of the sensitive information that your clients trust you with.
   (C) Using copyright to protect your source code is beneficial for the company because it provides a convenient way of securing intellectual property rights.
   (D) In theory, you as a software developer own the copyright to your software creation, and no one can copy, distribute, display, or make changes to it without your permission.

3. Which of the following is not right?
   (A) Technology is neutral.
   (B) The mindset of "Do it now, ask for forgiveness later" can be ruling business practices.
   (C) It is not the responsibility of software engineers to provide users with a secure and transparent program that they can trust.
   (D) All of the above.

Ⅲ. Fill in the numbered spaces with the words or phrases chosen from the box. Change the forms where necessary.

> speed   discover   enable   because   short
> focus   through   talk   only   process

CI/CD—One of the Latest Trends in Software Engineering

We really need to ___1___ about continuous integration and continuous deployment. CI/CD, for ___2___, this will be one of the most important software engineering trends in 2020 onwards.

A form of best practice, CI/CD enables IT, professionals, and service vendors, to streamline the software development ___3___ and improve their end-solution quality. Continuous integration, alone, is great as it allows specialists to ___4___ up application assemblies thanks to its instant error detections and code changes.

But, not ___5___ that, CI empowers professionals to put their code into a shared repository with each of their check-ins being verified. This process means bugs and errors, if any, are ___6___ in quick-time.

Continuous deployment is also an interesting development as it accelerates application update deliveries. This is ___7___ any changes to the core code are exposed to automated testing to assemble software builds for production ___8___ multiple deployments.

For these reasons, among others, CI/CD ___9___ teams to release apps and other software, very rapidly indeed. This kind of time-saving allows specialist team members to ___10___ on more complicated tasks and not get bogged down with the more bureaucratic elements of the software development life cycle.

Ⅳ. **Translate the following passage into Chinese.**

CMMI

The Capability Maturity Model Integration (CMMI) is a process and behavioral model that helps organizations streamline process improvement and encourage productive, efficient behaviors that decrease risks in software, product and service development.

The CMMI are continuous and staged models. Therefore, software organizations should choose one or other of the models, and also the disciplines that will be part of the model for the assessment and improvement of the software process.

Part 3

Simulated Writing: Business E-mail

E-mail is an electronic, computer-assisted online communication tool. In the business world it is used to transmit virtually every type of correspondence the daily conduct of business requires. Simple messages, memos and letters, complex reports, tables of data, graphs and charts, blueprints, pictures, you name it. If it can be generated by, scanned into, or downloaded onto a computer, it can be electronically sent through cyberspace to another computer.

In this guide you will read about writing business E-mails and helpful tips on formatting business E-mail. Each section provides useful information to assist you in becoming more proficient at using E-mail to communicate in the business world.

Formatting Business E-mail

The format of business E-mail is very similar to that of a business memo. So similar, in fact, that the basic heading elements found in a business memo are programmed into the computer generated template of every E-mail program. The communication role of business E-mail goes beyond that of a memo, however. In order not to overlook its versatility, the formatting elements of a business letter can be inserted manually into the body of an E-mail.

Business E-mail functions as both an internal and an external method of communication; its three main formatting elements are the heading, the body, and a signature block. Depending upon the nature of its correspondence and, at the discretion of the writer, business E-mail may also include a salutation and a complimentary close.

Business E-mail Heading

The heading of a business E-mail consists of up to six distinct information fields. They are located at the corner of the E-mail template, just below the tool bar.

The template itself appears automatically whenever you click on the New Mail, Reply, Reply to All, or the Forward button found on the tool bar of any E-mail program.

Each field in the template is designed to hold specific information, the definitions of which are preprinted on the left hand margin as follows:

- To:
- From:
- Cc:
- Bcc:
- Date:
- Subject:
- Attached:

Some of these fields are not always visible. The Bcc and Attached fields, for instance, are visible only when activated by the sender and, depending on the E-mail program, the From and Date fields may not be visible on the sender's template.

Business E-mail Body

The body of a business E-mail is no different than that of a business letter or memo.

The one formatting distinction is that E-mail programs automatically format the body in single spaced, full block style. It's a design function of the program and meant to ensure that the text of an E-mail appears on the recipient's screen exactly as it does on the sender's.

Short paragraphs are the rule, particularly as online readers often just scan the text. Many monitors display twenty to twenty-five lines at one time, making shorter paragraphs more suitable. Besides, long paragraphs are hard on the eyes and more difficult to read.

Always be considerate of your reader's time; an E-mail that goes on for more than two pages may be better off as a letter or memo. Delivered as an attachment, a lengthy letter or memo can be printed and read when time permits.

Generally speaking, business E-mail is sent in plain text rather than HTML. Using different font faces, colors, sizes, and styles, such as bold or italic, is extremely useful when creating documents within a word processor, but they are problematic when sent between E-mail programs.

Besides the possibility of increasing the download time on your recipient's computer, HTML documents may not appear on one screen as they do on another. If you are unfamiliar with the distinction and are offered a choice, just select plain text and move on.

Business E-mail Signature Block

The signature block in a business E-mail does the work of the heading or letterhead found in the format of a business letter. In other words, it supplies the contact information belonging to the sender. This is the last item in an E-mail. It is always located on the left hand margin below the signature line.

A signature block should contain all the contact information a recipient might require in order to respond to an E-mail. It should begin with the Sender's Name, Title, and Business Organization. A Physical Location, Phone Numbers, E-mail Address, and Web site should follow.

Business E-mail Salutations and Complimentary Closes

Including a salutation and complimentary close in a business E-mail is governed by the same rules as those governing business letters and memos.

Formal expressions such as Dear Ms. Ortiz and Sincerely yours are suitable for letter style business E-mails addressed to individuals with whom you are unfamiliar. When business E-mail functions as a memo, on the other hand, a salutation and complimentary close should be omitted altogether. More often than not, the salutation and complimentary close will be written informally, particularly after the protocols of an initial contact have been observed.

P. S. **This guide can be tailored according to the different business E-mails.**

A Sample Business E-mail

To: Expansion Community: itpartners@expansion.com, winpartners@expansion.com
From: Software Release Team
Cc: itag@expansion.com
Bcc: is&t@expansion.com
Date: Monday, August 23, 2021
Subject: SmartMessenger 15.0i Release Announcement
Attached: Brochure of SmartMessenger Enterprise 15.0i

Good Afternoon,

On behalf of the IS&T SmartMessenger Release Team, I am pleased to announce the release of SmartMessenger Enterprise 15.0i. SmartMessenger is an instant messaging program that runs on the Windows 10 client and Windows 2016 and 2019 server platforms. SmartMessenger Enterprise 15.0i contains many enhancements and new features, including the ability to support Android/iOS/PC and app payment.

How to Obtain:

Expansion Company possesses a site license for SmartMessenger, so the program is available to Expansion Company staff and affiliates free of charge. You can download the product from the Expansion Company IS&T Windows Software Site:
    https://web.expansion.edu/software/win.html

Please note that you need a current personal certificate to download this software. If you do not have a current personal certificate, then you can obtain one from https://www.expansion.com/.

Getting Help:
The SmartMessenger at Expansion Company page is located at:
http://itinfo.expansion.com/product.php?vid=644

If you have a question or need assistance, please contact the Computing Help Desk at: computing-help@expansion.com.

All the Best,

****
Kevin Li
SmartMessenger 15.0i Product Release Coordinator
Software Release Team

Unit 10 Beginning Your Work 开始工作

Information System Department
Expansion Company
Address: 12 Zhong Guancun Street, Haidian District, Beijing 100872, P. R. China
Phone: 8610-62310000
Fax: 8610-62311001
E-mail: kevinli@expansion.com
Website: http://www.expansion.com

 Exercises

There are two scenarios below, please choose one and write an E-mail.

Scenario 1: An E-mail to your manager about the work progress in the past week while let all of your team members know.

Scenario 2: An E-mail to your manager about your business trip plan next week.

Glossary(词汇表)

A

a host of		许多,一大群
a suite of		一系列,一套
acceptance testing		验收测试
accommodate	[əˈkɔmədeit] v.	考虑到,顾及
accommodate to		适应
account for		(在数量、比例方面)占
acronym	[ˈækrənim] n.	只取首字母的缩写词
act on		按照……而行动,对……起作用
actionable results		有目共睹的结果
actor	[ˈæktə(r)] n.	角色,行动者
ad hoc		特别的,专门的
adapt	[əˈdæpt] v.	适应
adaptive	[əˈdæptiv] adj.	适应的
add to it		加上
address	[əˈdres] v.	处理,满足,论述,重点提出
administer	[ədˈministə(r)] v.	管理,掌管
admiral	[ˈædmərəl] n.	海军上将,海军将官
advance	[ədˈvɑːns] v.	提升,提高
adverse	[ˈædvɜːs, ədˈvɜːs] adj.	不利的,相反的,敌对的
advocate	[ˈædvəkeit; -ət] v.	提倡,主张
aeronautics	[ˌeərəˈnɔːtiks] n.	航空学
aesthetical	[iːsˈθetikəl] adj.	美的,美学的
agent	[ˈeidʒənt] n.	使然力(指引起一定作用的人或其他因素)
agile methodologies		敏捷方法论,敏捷开发方法
agilist	[ˈædʒailist] n.	敏捷主义者,机敏者
agnostic	[ægˈnɔstik] adj.	不可知论的,怀疑的
ain't	[eint]	(= are not, am not, is not)不是
albeit	[ɔːlˈbiːit] conj.	虽然
algorithm	[ˈælgəriðəm] n.	算法
alike	[əˈlaik] adv.	同样地
all-round		全面的,多方面的,综合性的
alongside	[əˌlɔŋˈsaid] prep.	与……一起,与……同时
alpha testing		α测试

Glossary(词汇表)

among other things		其中
analogy	[əˈnælədʒi]n.	类比,类推
animation	[ˌæniˈmeiʃn]n.	动画
anticipate	[ænˈtisipeit]v.	预料,预见,预计(并做准备)
apply to		应用
appreciate	[əˈpriːʃieit]v.	赏识,欣赏,重视
approach	[əˈprəutʃ]v.	接洽
architectural design		体系结构设计,概要设计
argue	[ˈɑːgjuː]v.	主张,认为,表明,论证
arise	[əˈraiz]v.	出现,产生,上升
arise from		由……引起,起因于
artifact	[ˈɑːtifækt]n.	人工制品,手工艺品
as opposed to		与……截然相反,对照
as per		按照,依据,如同
assembly	[əˈsembli]n.	汇编
assimilate	[əˈsiməleit]v.	透彻理解,消化
associate	[əˈsəusieit, əˈsəusiət]n.	同事,伙伴
at first sight		乍一看,初看时
audit	[ˈɔːdit]n.	(质量或标准的)审查,检查
augmented	[ɔːgˈmentid]adj.	增强的
avert	[əˈvɜːt]v.	避免,防止
axiom	[ˈæksiəm]n.	公理,格言,规则

B

back-and-forth		反复地,来回地
backup	[ˈbækʌp]adj.	候补的,备份的
bare bone		基本框架,梗概
be attributed to		归因于
be concerned with		涉及,关于,关心,关注
be confronted with		面对
be derived from		源于,得自
be designated as		被指定为
be fraught with		充满,充满……的
be geared to do		准备好做
be grateful for		对……心存感激
be involved in		参与
be protected from		被保护以免受
be referred to		被提及,涉及
be suited for		适合于

be tied to		被束缚
bedrock	[ˈbedrɔk]n.	牢固基础,基本事实
beg	[beg]v.	请求,恳求
besides	[biˈsaidz]adv.	此外
best of all		首先,最好的是,最重要的是
beta testing		β测试
better	[ˈbetə(r)]v.	改善,胜过,超过
beyond	[biˈjɔnd]prep.	除……之外
bill	[bil]n.	账单,(餐馆的)账单
black-box testing		黑盒测试
blueprint	[ˈbluːprint]n.	蓝图,计划
board	[bɔːd]n.	董事会,理事会
bogeyman	[ˈbəugimæn]n.	可怕(或可憎)的人或物,妖怪,精灵
book	[buk]v.	登记,预订
boundary value analysis		边界值分析法
brain-twisting		费脑的
brainstorm	[ˈbreinstɔːm]v.	头脑风暴,集中各人智慧猛攻
branch to		分支到
break down		拆除,消除,破除(障碍或偏见)
breakup	[ˈbreikʌp]n.	分解
bring it to life		使……充满活力
bug	[bʌg]n.	程序缺陷,臭虫,漏洞
build	[bild]n.	编好的程序,构造
build in		使成为一部分,嵌入
building block		组成部分;构成要素
business scenario		业务场景,业务方案

C

cache	[kæʃ]v.	隐藏,缓存
call over		把……叫过来
capability	[ˌkeipəˈbiləti]n.	才能,能力
capture	[ˈkæptʃə(r)]v.	获得,捕获
cardinal	[ˈkɑːdinl]adj.	主要的,最重要的
catastrophically	[ˌkætəˈstrɔfikli]adv.	灾难性地
catch	[kætʃ]n.	隐藏的困难,暗藏的不利因素
categorical	[ˌkætəˈgɔrikl]adj.	分类的,按类别的
cater to		迎合,为……服务
cause-effect graphing		因果图法
cause to		导致,造成,带来

英文	音标	中文
cement	[siˈment] v.	加强，巩固（关系等）
center on		集中在，着重在
chaos	[ˈkeiɔs] n.	混乱
chart	[tʃɑːt] v.	详细计划，记录
check against		对照
check in		登记，记录，报到
check out		给……办理结账手续，核实
checkbox	[ˈtʃekbɔks] n.	复选框，检查框
cheer	[tʃiə(r)] n.	欢呼，喝彩
chief	[tʃiːf] n.	头目，首领
claim	[kleim] v.	要求（拥有），索取，索要
class	[klɑːs] v.	把……归入某等级，把……看作
class	[klɑːs] n.	类
code	[kəud] n.	道德准则，行为规范
code	[kəud] v.	编码
cognitive	[ˈkɔgnətiv] adj.	认知的，认识的
coherent	[kəuˈhiərənt] adj.	一致的，明了的，清晰的
cohesion	[kəuˈhiːʒn] n.	内聚
cohesiveness	[kəuˈhiːsivnəs] n.	内聚性，黏结性
color blind		色盲
color scheme		配色方案，色彩设计
come a long way		突飞猛进，取得很大进展
come across		偶遇，无意中发现
come out o		出自
come to		谈到，涉及
come to a conclusion		得出结论，告一段落
commence	[kəˈmens] v.	开始
comment	[ˈkɔment] n.	注释，注解
commit to		使（自己）致力于（某事或做某事）
commitment	[kəˈmitmənt] n.	提交，承诺
compact	[kəmˈpækt; ˈkɔmpækt] adj.	简明的，紧凑的
compare ... to		把……比作
compelling	[kəmˈpeliŋ] adj.	引人注目的，迷人的
competent	[ˈkɔmpitənt] adj.	有能力的，胜任的
compiler	[kəmˈpailə(r)] n.	编译器
compromise	[ˈkɔmprəmaiz] n.	妥协，折中
concurrency	[kənˈkʌrənsi] n.	并发性，并行性
conduct	[kənˈdʌkt] v.	实施，进行
configuration	[kənˌfigəˈreiʃn] n.	配置

conform to		遵守
consciously	[ˈkɔnʃəsli] adv.	有意识地
consensus	[kənˈsensəs] n.	（意见等）一致
conspire against		密谋反对
construct	[kənˈstrʌkt] n.	结构体
contemplate	[ˈkɔntempleit] v.	思考，预期
context	[ˈkɔntekst] n.	上下文，环境，背景
contextual	[kənˈtekstʃuəl] adj.	与上下文有关的，与语境相关的
control	[kənˈtrəul] n.	控件
converge on		集中于
convergence	[kənˈvɜːdʒəns] n.	一体化，集中，收敛
convey	[kənˈvei] v.	传达
cool	[kuːl] adj.	棒的，酷的
cope with		对付，应付
correspondence	[ˌkɔrəˈspɔndəns] n.	通信
cost-effective		有成本效益的，划算的
couple	[ˈkʌpl] v.	耦合
course	[kɔːs] n.	过程，进程
craft	[krɑːft] v.	精巧地制作
criterion	[kraiˈtiəriən] n.	标准，准则（复数为 criteria）
cross-functional		跨职能的
cross-reference		（书等）前后对照
culminate	[ˈkʌlmineit] v.	告终，达到顶点
cultivate	[ˈkʌltɪveit] v.	培养，陶冶
curate	[ˈkjuərət] v.	操持（收藏品或展品的）展出，组织（音乐节的）演出
cut corners		抄近路，以简便方法做事
cut shot		缩短，打断，缩减
cutting-edge		先进的

D

daunt	[dɔːnt] v.	沮丧，使气馁
deal with		处理
debug	[ˌdiːˈbʌg] v.	调试
decommission	[ˌdiːkəˈmiʃn] v.	停止使用，使退役
decomposition	[ˌdiːkɔmpəˈziʃn] n.	分解
dedicate to		把（时间、精力等）用于
deduce	[diˈdjuːs] v.	推论，演绎出
deem	[diːm] v.	认为，视作

default to		默认为
defect	[diːfekt; diˈfekt]n.	缺陷
definitive	[diˈfinətiv]adj.	最后的,决定性的
degrade	[diˈgreid]v.	退化,削弱(尤指质量)
delegate	[ˈdeligət, ˈdeligeit]v.	委托,授(权),把……委托给他人
deliverable	[diˈlivərəbl]n.	应交付的产品
delivery	[diˈlivəri]n.	交付
demonstrate	[ˈdemənstreit]v.	证明,论证
depict	[diˈpikt]v.	描述,描写
deployment	[diˈplɔimənt]n.	调度,部署
deployment pipeline		部署流水线
deposit	[diˈpɔzit]n.	押金,预付金
deserve	[diˈzɜːv]v.	值得,应得
desirable	[diˈzaiərəbl]adj.	可取的,值得做的
desire	[diˈzaiə(r)]n. & v.	愿望,欲望,渴望
desk-check		手工检查
destabilize	[ˌdiːˈsteibəlaiz]v.	使不安定,使不稳定
detailed design		详细设计
development bootcamp		开发训练营
die	[dai]v.	停止运转
dilemma	[diˈlemə]n.	(进退两难的)窘境,困境
discipline	[ˈdisəplin]n.	学科,方法
discipline	[ˈdisəpln]v.	自我控制,严格要求(自己)
discrete	[diˈskriːt]adj.	离散的,不连续的
disseminate	[diˈsemineit]v.	传播(信息、知识等)
dissolve	[diˈzɔlv]v.	解散
distinct	[diˈstiŋkt]adj.	不同的,有区别的
distinctive	[diˈstiŋktiv]adj.	区别性的,特殊的
distribute	[diˈstribjuːt]v.	分销
distribution	[distribjuːʃn]n.	经销,分销
divide-and-conquer		分而治之
do one's bit		尽自己的一份力量
document	[ˈdɔkjumənt]v.	记录,纪实性地描述
doubtless	[ˈdautləs]adv.	大概,几乎肯定地
draft	[drɑːft]v.	起草,制定
drastically	[ˈdræstikli; ˈdrɑːstikli]adv.	彻底地
drill-down		深度探讨
drop-down		下拉菜单

E

edge	[edʒ] n.	优势
edge case		边界用例,极端例子
egoless	[ˈiːɡəulis] adj.	不自负的,非自我的
elaboration	[iˌlæbəˈreiʃn] n.	精化
elevate	[ˈeliveit] v.	提高,提升
elicit	[iˈlisit] v.	得出,诱出,引起
elicitation	[iˌlisiˈteiʃn] n.	导出,启发
elusive	[iˈluːsiv] adj.	难以被发现的
embed	[imˈbed] v.	嵌入
embellishment	[imˈbeliʃmənt] n.	修饰,润色
embrace	[imˈbreis] v.	掌握,接受,拥抱
employ	[imˈplɔi] v.	使用,应用
empower	[imˈpauə(r)] v.	使能够,授权
encapsulation	[inˌkæpsjuˈleiʃn] n.	封装
encompass	[inˈkʌmpəs] v.	包含,包括
end up		结束
end-stage		末期的
end-to-end		端到端
endeavor	[inˈdevə] n.	努力,尽力
energetic	[ˌenəˈdʒetik] adj.	精力充沛的
engage	[inˈɡeidʒ] v.	吸引住,引起注意
engagement	[inˈɡeidʒmənt] n.	参与度
enumeration	[iˌnjuːməˈreiʃn] n.	列举事实,逐条陈述
epsilon	[ˈepsilɔn, epˈsailən] n.	小(或近于零)的正数
equivalence class partitioning		等价类划分法
essence	[ˈesns] n.	本质,实质
ever-changing		千变万化的,常变的
every now and then		不时地,常常
evolutionary	[ˌiːvəˈluːʃənri] adj.	逐渐发展的,演变的
except for		除……之外
exercise	[ˈeksəsaiz] v.	执行,使用
exert	[iɡˈzəːt] v.	施以影响,施加(压力等)
exhaustive	[iɡˈzɔːstiv] adj.	详尽的,彻底的,全面的
exploitable	[iksˈplɔitəbl] adj.	可利用的,有利可图的
eXtreme Programming		极限编程

F

facet	['fæsit] n.	面,方面
facilitate	[fə'siliteit] v.	使容易,使便利,促进
fairly	['feəli] adv.	相当地；公平合理地
fall of		脱落,掉队
fall under		归入……类,属于
famine	['fæmin] n.	饥荒,奇缺
fan-in		扇入,输入端
fan-out		扇出,输出端
faulty	['fɔːlti] adj.	有错误的；有缺陷的
features midstream		中游特色
fictitious	[fik'tiʃəs] adj.	虚构的,假想的,编造的
figure out		想出,理解,断定
finance	['fainæns] v.	供给……经费,负担经费
financial statement		财务报表
fine	[fain] adv.	够好,蛮不错
finger-pointing		指责,责难
finite	['fainait] adj.	有限的,限定的
fit for		适合
fit in with		适应,符合,与……一致
fix	[fiks] n.	仓促的解决办法
fix	[fiks] v.	确定,处理,解决
fixture	['fikstʃə(r)] n.	固定装置
flip	[flip] v.	翻转,快速翻动
flowchart	['fləutʃɑːt] n.	流程图
fluid	['fluːid] adj.	变化的,流动的,液体的
for compliance with		遵守,遵从,符合
for short		作为简称,作为缩写
formal and rigorous methods		形式化和精确方法
foster	['fɔstə(r)] v.	促进
framework	['freimwɜːk] n.	构架,体系结构,准则
freakish	['friːkiʃ] adj.	反常的,奇怪的
free from		免于,使摆脱
free of		没有……的
free ride offer		免费乘车优惠
free-form		自由形态的,结构不规则的
from end to end		从头到尾地
from scratch		从头做起,白手起家

fruitless	[ˈfruːtləs] *adj.*	徒劳的,无用的
fulfill	[fulˈfil] *v.*	实现
fully-fledged		完善的,成熟的,羽毛丰满的
function	[ˈfʌŋkʃn] *n.*	函数
functioning	[ˈfʌŋkʃəniŋ] *n.*	功能
funding	[ˈfʌndiŋ] *n.*	提供资金

G

Gantt chart		甘特图
gap analysis		差异分析
general-purpose		通用的
generally speaking		一般说来
generation	[ˌdʒenəˈreiʃn] *n.*	一代,阶段
generic	[dʒəˈnerik] *adj.*	一般的,普通的
genius	[ˈdʒiːniəs] *n.*	天才,天才人物
get along with		与……和睦相处,取得进展
get involved in		涉及,卷入
get one's hands on		得到,占有
get squeezed to		陷入,挤到
get used to		习惯
given	[ˈgivn] *n.*	假设
go about		着手做
go on and on		一直继续下去,说个不停
go through		经受,仔细检查
go under		归为(某一类事物)中
go way up		大幅上升
grey-box testing		灰盒测试

H

hand-in-hand		手牵手的,亲密的,并进的
hands-on		亲身实践的,实际动手做的
head	[hed] *v.*	领导,主管
headquarter	[ˌhedˈkwɔːtə] *v.*	总部设于……,总公司设于……
heaviest	[ˈhevist] *adj.*	(在数量、程度等方面)超出一般的
heuristics	[hjuˈristiks] *n.*	启发法,启发式教学法
hierarchical	[ˌhaiəˈrɑːkikl] *adj.*	分等级的,层次的
high season		旺季
highlight	[ˈhailait] *v.*	强调,使突出
hitherto	[ˌhiðəˈtuː] *adv.*	迄今,至今

holistic	[həˈlistik] *adj.*	整体的,全面的,功能整体性的
homogeneous	[ˌhəməˈdʒiːniəs] *n.*	相同特征的,同性质的
house	[haus] *v.*	安置,存放
humongous	[hjuːˈmʌŋgəs] *adj.*	巨大无比的,极大的
hungry for		渴望

identify	[aiˈdentifai] *v.*	确认,说明身份
identify with		视……为一体,认同
if only		要是……多好,只要
imperative	[imˈperətiv] *adj.*	重要的,必要的
implicitly	[imˈplisitli] *adv.*	暗含地,含蓄地
impose	[imˈpəuz] *v.*	强制实行,把……强加于
in a nutshell		极其简括地说
in all honesty		老实说
in and of itself		就其本身而言
inception	[inˈsepʃn] *n.*	开端
in charge of		负责,领导
in detail		详细地
in exchange for		作为交换
in favor of		支持,赞同
in haste		急忙地,草率地,慌张地
in isolation of		脱离
in large part		大体上,在很大程度上
in moderation		适中地
in part		在某种程度上,部分地
in place		适当,就位,(法律、政策、行政体系等)正在运作
in reaction to		应对
in-depth		深入详尽的,彻底的
in-house		内部的
incentive	[inˈsentiv] *n.*	激励,刺激
incorporate	[inˈkɔːpəreit] *v.*	加入,包含
incubation	[ˌiŋkjuˈbeʃn] *n.*	培育
indispensable	[ˌindiˈspensəbl] *adj.*	不可缺少的,必需的
industry vertical		产业垂直
inflated cost		滥计成本,虚列成本
inform of		将……告知(某人)
ingredient	[inˈgriːdiənt] *n.*	因素,成分

inheritance	[in'herit(ə)ns]n.	继承
inhibit	[in'hibit]v.	阻止,阻碍
insight	['insait]n.	洞察力
instantaneous	[ˌinstən'teɪniəs]adj.	立即的,立刻的,瞬间的
instill	[in'stil]v.	逐渐灌输
institution	[ˌinsti'tjuːʃn]n.	习俗,制度
instruction	[in'strʌkʃn]n.	指令
integrated circuits		集成电路
integration testing		集成测试
interest	['intrəst,'intrest]v.	使……感兴趣,使……参与
internship	['intɜːnʃip]n.	实习生职位,实习期
interpreter	[in'tɜːprətə(r)]n.	解释器,解释程序
intimate	['intimət,'intimeit]adj.	精通的,基本的
intrinsic	[in'trinzik]adj.	固有的,内在的,本身的
introduce	[ˌintrə'djuːs]v.	引入
introspection	[ˌintrə'spekʃn]n.	内省,反省
intuitive	[in'tjuːitiv]adj.	直觉的,凭直觉获知的
invoke	[in'vəuk]v.	调用
issue	['iʃuː]v.	发出,(正式)发给
item	['aitəm]n.	项
iterative	['itərətiv]adj.	迭代的,重复的,反复的

J

jaw-breaking		难发音的,读起来费劲的
jell	[dʒel]n. & v.	结合,相处融洽
judge	[dʒʌdʒ]v.	判断,评价
juggernaut	['dʒʌgənɔːt]n.	巨无霸,世界主宰,强大的破坏力
just about		几乎,差不多

K

key takeaway		关键点,关键信息,重要信息
knit	[nit]v.	结合

L

labor intensive		劳动密集型,人工密集
late-in-the-game		比赛后期
laundry-list		细目清单
law	[lɔː]n.	规则,法则
lay out		划定(路线),布置,安排

lead to		导致
letter	[ˈletə(r)]n.	字母
leverage	[ˈliːvəridʒ]v.	施加影响,利用
literally	[ˈlitərəli]adv.	真实地,确切地
literature	[ˈlitrətʃə(r)]n.	著作,文献
live	[laiv]adj.	生动的
locate	[ləuˈkeit]v.	找出……的准确位置
log	[lɔg]n.	日志
log	[lɔg]v.	记录
long-range		长远的,长期的
look into		调查,窥视,观察
loop	[luːp]n.	循环
loophole	[ˈluːphəul]n.	(法律、合同等的)漏洞
low season		淡季

M

maintainer	[meinˈteinə]n.	软件维护人员
major	[ˈmeidʒə(r)]adj.	较大的,较重要的
management	[ˈmænidʒmənt]n.	经营者,管理
managerially	[ˌmænəˈdʒiəriəli]adv.	管理地
manifest	[ˈmænifest]v.	显示,表明
manifestation	[ˌmænifeˈsteiʃn]n.	显示,表明,表示
manifesto	[ˌmæniˈfestəu]n.	宣言,声明
manipulate	[məˈnipjuleit]v.	控制,操作,使用
map out		设计,规划,安排
margin	[ˈmɑːdʒin]n.	利润
markedly	[ˈmɑːkidli]adv.	显著地,明显地
marketability	[ˌmɑːkitəˈbiləti]n.	可销售性,可流通性
markup	[ˈmɑːkʌp]n.	标记
master	[ˈmɑːstə(r)]adj.	支配的,总的,最重要的
mechanism	[ˈmekənizəm]n.	机制
mental	[ˈmentl]adj.	思考的,智力的,脑力的
meta-information		元信息
metalanguage	[ˈmetəlæŋgwidʒ]n.	元语言
metaphorical	[ˌmetəˈfɔrikl]adj.	隐喻性的,比喻性的
milestone	[ˈmailstəun]n.	里程碑
mindset	[ˈmaindset]n.	观念模式,思维倾向
minimalist	[ˈminiməlist]n.	最低限要求者
minor	[ˈmainə(r)]adj.	较小的,次要的

mitigation	[ˌmiti'geiʃn] n.	缓解,减轻
mockup	['mɔkʌp] n.	实物模型
model	['mɔdl] v.	建立模型
modularity	[ˌmɔdju'læriti] n.	模块性
momentum	[mə'mentəm] n.	势头,动力,冲力
morale	[mə'rɑ:l] n.	士气,斗志
motivate	['məutiveit] v.	激发……的积极性
multiply	['mʌltiplai] v.	使增加
mutant	['mju:tənt] n.	变体
mutation testing		变异测试,突变测试
mythical	['miθikl] adj.	神话的,虚构的

N

namely	['neimli] adv.	即,也就是
narrative	['nærətiv] n. & adj.	叙述,故事,讲述的
next in line		(按顺序的)下一个
no point		毫无意义,没理由,无济于事
not yet		还没有
notable	['nəutəbl] adj.	著名的
notably	['nəutəbli] adv.	尤其,特别,极大程度上
notation	[nəu'teiʃn] n.	符号
note	[nəut] v.	记录
notion	['nəuʃn] n.	概念,观念,看法

O

object	['ɔbdʒikt] n.	对象,目标
obviate	['ɔbvieit] v.	排除,避免
often-cited		经常被引用的
omission	[ə'miʃn] n.	疏忽,遗漏
on the contrary		正相反
on this aspect		在这个方面
operation	[ˌɔpə'reiʃn] n.	运营,经营
opportune	['ɔpətju:n] adj.	恰好的,适当的
order of magnitude		数量级
outsource	['autsɔ:s] v.	外包
over budget		超过预算
overdue	[ˌəuvə'dju:] adj.	过期的,迟到的
overriding	[ˌəuvə'raidiŋ] adj.	首要的,压倒一切的
oversee	[ˌəuvə'si:] v.	管理,监督

oversight	[ˈəuvəsait]n.	疏忽,漏失

P

pacemaker	[ˈpeismeikə(r)]n.	起搏器
package	[ˈpækidʒ]v.	包装(商品)
pain-point		痛点
pair	[peə(r)]v.	把……组成一对
pair	[peə(r)]v.	使成对,配对
pane	[pein]n.	(一片)窗玻璃
paper	[ˈpeipə(r)]n.	论文,文章
partition	[pɑːˈtiʃn]v.	划分,把……分成部分
paycheck	[ˈpeitʃek]n.	薪金,工资,(支付薪金或工资的)支票
peace of mind		内心的宁静,宁静如水
peculiarly	[piˈkjuːliəli]adv.	特有地,特别地
penetration testing		渗透测试,渗透检查
perfect	[ˈpɜːfikt, pəˈfekt]v.	使完善,改进
performance	[pəˈfɔːməns]n.	性能,业绩,工作情况
performance testing		性能测试
periodic	[ˌpiəriˈɔdik]adj.	定期的,周期的
phew	[fjuː]int.	哎呀,唷(表示不快、惊讶的声音)
philosophy	[fəˈlɔsəfi]n	方法
place an order		订购,下单,发出订单
place on		寄托,把……放在……上
play by the rules		遵守规则,按规则出牌
please	[pliːz]v.	使喜欢,使高兴,使满意
plug	[plʌg]n.	插头,插座
plug-in		插件,外挂程序
pointing fingers		指手画脚
polish up		改善,润色,使完美
polymorphism	[ˌpɔliˈmɔːfizm]n.	多态
pool	[puːl]n.	资源的集合,联营
pop into one's mind		突然想到
populate	[ˈpɔpjuleit]v.	构成,填充
pose	[pəuz]v.	引起,造成
post	[pəust]n.	岗位
post-condition		后置条件
power	[ˈpauə(r)]v.	促进,推动,驱动
practice	[ˈpræktis]n.	实践,通常的做法,惯例
practitioner	[prækˈtiʃənə(r)]n.	从业者,实践者

pre-condition		前置条件,先决条件
precept	[ˈpri:sept]n.	规则
predate	[ˌpri:ˈdeit]v.	(在日期上)早于,先于
predatory	[ˈpredətri]adj.	掠夺性的,欺负弱小的,压榨他人的
premier	[ˈpremiə(r)]adj.	首要的,最著名的
prescribe	[priˈskraib]v.	规定,命令,指示
presence	[ˈprezns]n.	势力
price	[prais]v.	给……定价,(在商品上)标价
primarily	[praiˈmerəli]adv.	原来,根本上
principally	[ˈprinsəpli]adv.	主要地,大部分地
print	[print]n.	印刷行业,出版界
prioritize	[praiˈɔrətaiz]v.	把……区分优先次序
proceed	[prəˈsi:d]v.	继续进行
proficient	[prəˈfiʃnt]adj.	娴熟的,精通的,训练有素的
profile	[ˈprəufail]v.	描……的轮廓,扼要描述
profiler	[ˈprəufailə(r)]n.	曲线图,概览图,分析工具
profound	[prəˈfaund]adj.	意义深远的,深刻的,深奥的
prominent	[ˈprɔminənt]adj.	著名的
promotion	[prəˈməuʃn]n.	(商品等的)宣传,推销(活动)
proponent	[prəˈpəunənt]n.	支持者,倡导者
proposition	[ˌprɔpəˈziʃn]n.	主张,提议,建议
proprietor	[prəˈpraiətə(r)]n.	业主,所有人
protocol	[ˈprəutəkɔl]n.	(数据传递的)协议,规程
prototype	[ˈprəutətaip]n.	原型
provocative	[prəˈvɔkətiv]n.	引起争论(议论,兴趣等)的
pseudocode	[ˈsju:dəukəud]n.	伪代码
pursue	[pəˈsju:]v.	从事,进行
put into use		应用,使用
put off		推迟
put through		完成
puzzle	[ˈpʌzl]n.	不解之谜,疑问
pyramid	[ˈpirəmid]n.	金字塔形的物体(或一堆东西),金字塔

R

rank	[ræŋk]n.	等级
rate	[reit]v.	评估,评价
read through		通读
recipe	[ˈresəpi]n.	诀窍
recognize	[ˈrekəgnaiz]v.	认可,承认

recur	[rɪˈkɜː(r)] v.	循环
refer to		提到,参考,查阅
reference	[ˈrefərəns] v.	把……引作参考,引用
reflection	[rɪˈflekʃ(ə)n] n.	反射
refund	[ˈriːfʌnd] v.	退还,偿还
regime	[reɪˈʒiːm] n.	体制,体系
regression testing		回归测试
rehash	[ˈriːhæʃ] v.	事后反复回想(或讨论)
release	[rɪˈliːs] v.	发布,公开发行
reminder	[rɪˈmaɪndə(r)] n.	提醒,提示
repository	[rɪˈpɒzətri] n.	知识库,仓库,存储库
resemble	[rɪˈzembl] v.	看起来像,显得像
respective	[rɪˈspektɪv] adj.	分别的,各自的
restock	[ˌriːˈstɒk] v.	补充(货源),再补给
result from		由……引起
resultant	[rɪˈzʌltənt] adj.	因此而产生的
retract	[rɪˈtrækt] v.	撤回,收回(协议、承诺等)
reuse	[ˌriːˈjuːz] n.	复用
revision	[rɪˈvɪʒn] n.	修改,修订
revolve around		围绕着
rich	[rɪtʃ] adj.	宝贵的
right	[raɪt] adv.	正确地,恰当地,彻底地
rightly	[ˈraɪtli] adv.	确实地
ring out		(根据考勤钟)打卡下班
ripple back to		回溯到
road map		线路图,(一步一步的)详尽计划
rollout	[ˈrəʊlaʊt] n.	首次展示,(产品或服务的)正式推出
room	[ruːm] n.	余地,空间
rotate	[rəʊˈteɪt] v.	轮流,交替
rough	[rʌf] adj.	初步的,粗略的
routine	[ruːˈtiːn] adj. n.	常规的,日常的 程序,例行程序
rule of thumb		单凭经验的方法

S

sanction	[ˈsæŋkʃn] v.	批准,认可,同意
saying	[ˈseɪɪŋ] n.	谚语,格言
scale	[skeɪl] v.	按比例决定,测量
scale down		按比例减少,按比例缩小

英文	音标	中文
scenario	[səˈnɑːriəu] n.	场景,某一特定情节
schematic	[skiːˈmætik] adj.	示意性的,图示的
scrap	[skræp] v.	废弃,取消
script	[ˈskript] n.	脚本
self-contained		独立的
self-healing		自愈
semicolon	[ˌsemiˈkəulən] n.	分号
serve as		充当,用作
session	[ˈseʃn] n.	(议会等的)会议
settle on		选定,决定
shady	[ˈʃeidi] adj.	非法的
sheer	[ʃiə(r)] adj.	纯粹的,绝对的
shorten	[ˈʃɔːtn] v.	减少
sick leave		(尤指拿病假工资的)病假
side effect		副作用,意外的连带后果
sift	[sift] v.	详审,精选
silo	[ˈsailəu] v. & n.	将……贮藏在筒仓中,筒仓,数据仓库
sizable	[ˈsaizəbl] adj.	相当大的,很大的
sketch out		草拟,概略地叙述
sloppy	[ˈslɔpi] adj.	草率的,粗心的
smoke test		冒烟测试
smoothly	[ˈsmuːðli] adv.	平稳地,连续而流畅地,顺利地
sophomore	[ˈsɔfəmɔː(r)] n.	大学二年级学生
spawn	[spɔːn] v.	大量产生,造成,引发,引起
specification	[ˌspesifiˈkeiʃn] n.	说明书,规范
specify	[ˈspesifai] v.	详细说明,指定
sphere	[sfiə(r)] n.	范围
spirited	[ˈspiritid] adj.	热烈的
spoil	[spɔil] v.	破坏,糟蹋
stack	[stæk] n.	一堆,大量,许多
staff	[stɑːf] v.	配置职员
stakeholder	[ˈsteikhəuldə(r)] n.	利益相关者,干系者
stand for		表示,代表
stand in the way		碍事,挡道
start-up		启动阶段的,开始阶段的
state-based testing		基于状态的测试
state	[steit] v.	陈述,说明
statement	[ˈsteitmənt] n.	语句
stem from		基于,处于

step back		后退，退后
step in		介入，插手干预
stepwise	['stepwaiz]adj.	逐步的
stereotype	['steriətap]n.	模式化观念(或形象)，老一套，刻板印象
stress	[stres]n.	压力，紧迫
stress testing		压力测试
stringent	['strindʒənt]adj.	严格的，严厉的
subclass	['sʌbklɑ:s]n.	子类
subjective	[səb'dʒektiv]adj.	主观的，个人的
subordinate	[sə'bɔ:dinət, sə'bɔ:dineit]n.	下级，部属
subsequent	['sʌbsik(ə)nt]adj.	随后的，继……之后的
suffer	['sʌfə(r)]v.	受损失，受害
suffice	[sə'fais]v.	足够，满足
suit	[su:t; sju:t]v.	适合，合乎……的要求
sum up		总结，概述
superior	[su:'piəriə(r), sju:'pəriə(r)]n.	上级，高手
surface	['sɜ:fis]v.	出现，显露
swipe	[swaip]v.	滑动屏幕
symptomatic	[ˌsimptə'mætik]adj.	有症状的
synchronization	[ˌsiŋkrənai'zeiʃn]n.	同步
synchronously	['siŋkrənəsli]adv.	同时地，同步地

T

tabular	['tæbjələ(r)]adj.	列成表格的
tackle	['tækl]v.	处理，解决
tailor	['teilə(r)]v.	剪裁，适应
take advantage of		利用
take charge of		担任，监管
take the lead		带头，一马当先
take up		占(空间、时间)
takes a view of		观察，检查
talk over		商议，讨论
temperament	['temprəmənt]n.	气质，性情，性格
test case		测试用例
text box		文本框
that being said		尽管如此，另一方面
that is to say		即，换句话说

the point is		问题在于
thinker	[ˈθiŋkə(r)]n.	思想者
thread	[θred]n.	各组成部分
thrice	[θrais]n.	三次,三倍
through to		直到,一直到
throw up		突然建造,匆忙建造
timely	[ˈtaimli]adj.	及时的,适时的
to this end		为了这个目的,为了达到这个目标
trade-off		折中,平衡
traditionalist	[trəˈdiʃənəlist]n.	传统主义者,墨守成规者
trial use		试用,试行
trait	[treit]n.	特性
tricky	[ˈtriki]adj.	难办的,难对付的
trivial	[ˈtriviəl]adj.	不重要的,价值不高的,微不足道的
troubleshoot	[ˈtrʌblʃu:t]v.	解决重大问题,排除……的故障
trust sb. with sth./sb.		托付,托交,把……委托给
try out		试验,尝试
tumble	[ˈtʌmbl]v.	使倒下,搅乱
turn away from		回避
typographical	[ˌtaipəˈgræfikl]adj.	印刷上的,排字上的
typography	[taiˈpɔgrəfi]n.	排版

U

ultimately	[ˈʌltimətli]adv.	最后,终于,根本
uncover	[ʌnˈkʌvə(r)]v.	发现,揭开,揭露
under the radar		避开别人关注的行为,低调处理
under the supervision of		在……的监督下,在……的指导下
underlie	[ˈʌndəˈlai]v.	构成……的基础,位于……之下
undertake	[ˌʌndəˈteik]v.	着手做,开始进行
unearth	[ʌnˈɜ:θ]v.	发现,揭露
unit testing		单元测试
up to		(时间上)一直到
use case		用例
utility	[ju:ˈtiləti]n.	公用事业,公共事业设备

V

validation	[ˌvæliˈdeiʃən]n.	验证,确认
validation testing		确认测试
value	[ˈvælju:]v.	评价

value proposition		价值主张，价值定位
variant	['veəriənt]n.	变体，转化
verification	[ˌverifi'keiʃn]n.	验证，确认，验收
versus	['vɜːsəs]prep.	（比较两种不同想法、选择等）与……相对，与……相比
view	[vjuː]n.	视图，观点
virtually	['vɜːtʃuəli]adv.	事实上，实质上
vision	['viʒn]n.	愿景，视野，想象
visual	['viʒuəl]n.	视觉资料（指说明性的图片、影片等）
vital	['vaitl]adj.	至关重要的，生死攸关的
volatile	['vɔlətail]adj.	不稳定的，易变的
vulnerability	[ˌvʌlnərə'biləti]n.	弱点，脆弱性

W

watch out for		密切注意，戒备，提防
wear out		用坏，磨损
well done		做得好
well-defined		明确的
well-led		领导得好的
well-oiled		（系统或机构）运转顺畅的，运行良好的
while	[wail]conj.	虽然，尽管
white-box testing		白盒测试
wireframe	['waiəfreim]n.	示意图，线框模型
with…in mind		把……放在心上，以……为目的
with a view to		为了，目的在于
with access to		访问
with an eye to		着眼于
with respect to		就……而论，关于
with the advent of		随着……的出现
with the help of		借助
withhold	[wið'həuld]v.	拒绝给，不给
word of mouth		口碑，口头传述的
work out		解决，实现
work the way		进行
work through		解决，完成，干完
workload	['wɜːkləud]n.	工作量
write up		详细写出

Abbreviation(缩略语表)

A

ACM	Association for Computing Machinery	国际计算机学会
API	Application Programming Interface	应用程序接口
ARPANET	Advanced Research Projects Agency Network	美国高级研究计划署网络

B

BPO	Business Process Outsourcing	业务流程外包

C

CBDMO	Chief Business Development and Marketing Officer	首席商业扩展和营销官
CEO	Chief Executive Officer	首席执行官,执行总裁
CFO	Chief Financial Officer	首席财务官,财务总监
CIO	Chief Information Officer	首席信息官
COO	Chief Operating Officer	首席运营官
CSO	Chief Strategy Officer	首席策略官
CTO	Chief Technology Officer	首席技术官

D

DARPA	Defense Advanced Research Projects Agency	美国国防部高级研究计划署
DFD	Data Flow Diagram	数据流图

E

EIA	Electronic Industries Association	美国电子工业协会

G

GUI	Graphical User Interface	图形用户界面

I

IC	Integrated Circuit	集成电路,芯片
IEEE	Institute of Electrical and Electronics Engineers	美国电气和电子工程师协会
IFIPS	International Federation of Information	国际信息处理学会联合会

Processing Societies

N

NATO	North Atlantic Treaty Organization	北大西洋公约组织

O

OO	Object-Oriented	面向对象的

R

R&D	Research and Development	研究与开发

S

SCM	Software Configuration Management	软件配置管理
SDK	Software Development Kit	软件开发工具包
SGML	Standard for General Markup Language	通用标记语言标准

U

UI	User Interface	用户界面
UML	Unified Modeling Language	统一建模语言

V

V&V	Verification and Validation	验证和确认
VIP	Very Important Person	重要人物,大人物

W

W5HH	Why, What, When, Who, Where, How, How much	为什么,什么,什么时候,谁,在哪里,如何,有多少

Answers(练习答案)

Unit 1

Part 1 Listening and Speaking

Listening Comprehension

1. (C) 2. (B) 3. (A)

原文如下：

Software Engineering

Today, computer software has become the key element in the evolution of computer-based systems and products and one of the most important technologies on the world stage. Over the past 70 years, software has evolved from a specialized problem solving and information analysis tool to an industry in itself and becomes a dominant factor in the economies of the industrialized world.

Yet we still have trouble developing high-quality software on time and within budget. Why does it take so long to get software finished? Why are development costs so high? Why can't we find all errors before we give the software to our customers? Why do we spend so much time and effort maintaining existing programs? And why do we continue to have difficulty in measuring progress as software being developed and maintained? These questions and many others **demonstrate** the industry's concern about software and the manner in which it is developed — a concern that has lead to the adoption of software engineering **practice**.

Then, what is software? And which characteristics of software make it different from other things that human beings build? Software is a logical rather than a physical system element, and some characteristics make software special. First, it is difficult for a customer to specify requirements completely, and difficult for the supplier to understand fully the customer needs as well. In defining and understanding requirements, especially changing requirements, large quantities of information need to be communicated and **assimilated** continuously. Second, software is seemingly easy to change and is **primarily** intangible, and much of the process of creating software is also intangible, involving experience, thought and imagination. In addition, it is difficult to test software **exhaustively**.

Software — programs, data, and documents — **addresses** a wide range of technology and application areas, yet all software evolves according to a set of **laws** that have remained the same for over 50 years. The intent of software engineering is to provide a **framework** and a solution for building higher quality software. It includes greater emphasis on systematic development, a concentration on finding out the user's requirements, formal/semi formal specifications of the requirements of a system, demonstration of early version of a system (**prototype**), greater emphases on trying to ensure error free code and so on.

From the time in 1968 when the phrase "software engineering" was first used at a **NATO** conference until the present day, software has come a long way. But it still has a very long way to go if it is to be considered as mature as other engineering **disciplines**.

Dictation

1. known   2. project manager   3. Operating System   4. human   5. publication
6. bookshelf   7. statement   8. manpower   9. later   10. law
11. environments 12. collaborative 13. individuals   14. originally   15. embodied

16. drop   17. circulation   18. programming   19. productivity   20. stimulated

Part 2　Reading and Translating

Section A

Ⅰ. 1. F　2. F　3. T　4. T　5. F
Ⅱ. 1.（D）　2.（D）　3.（B）
Ⅲ.
1. begins　2. involved　3. developers　4. refers　5. connecting
6. using　7. behind　8. collaborate　9. needed　10. cycle

Ⅳ. 软件演化

在软件工程领域，软件演化是指基于各种原因的软件开发、软件维护以及软件更新的过程。软件改变是不可避免的，因为在一个软件的生命周期中有许多因素会发生改变。这些因素包括：

- 需求的变更
- 环境的变化
- 错误或安全漏洞
- 添加了新设备或删除了旧设备
- 系统的改进

对于很多公司，它们最大的业务投资之一就是软件和软件开发。软件被视作是非常关键的资产，因此管理者希望确保他们所雇佣的软件工程师团队能够致力于确保软件系统通过不断演化来保持最新。

Section B

Ⅰ. 1. F　2. F　3. T　4. T　5. F
Ⅱ. 1.（C）　2.（D）　3.（D）
Ⅲ.
1. going　2. level　3. aiming　4. create　5. leaders
6. corner　7. core　8. out　9. evolving　10. learned

Ⅳ. 基于 AI 的复杂软件项目管理方法

人工智能（AI）对经济、行政和社会等各个领域都有十分重大的影响。AI 在软件工程领域有一个意想不到的应用：AI 首次为软件开发提供了可靠的方法，以便分析和评估复杂的软件及其开发过程。存储库挖掘、机器学习、大数据分析和软件可视化都可以为软件质量、软件开发和软件项目管理提供有针对性的见解和有效的预测。

Unit 2

Part 1　Listening and Speaking

Listening Comprehension

1.（A）　2.（C）　3.（C）

原文如下：

Software Requirements

It is necessary to understand requirements before design and construction of a computer-based system can begin. To accomplish this, a set of requirements engineering tasks are conducted. Requirements engineering occurs during the customer communication and modeling activities that we have defined for the generic software process. Seven distinct requirements engineering functions — inception, elicitation, elaboration, negotiation, specification, validation, and management — are conducted by members of the software team.

At project inception, the developer and the customer as well as other stakeholders establish basic problem requirements, define overriding project constraints, and address major features and functions that must be present for the system to meet its objectives. This information is refined and expanded during elicitation, which is a requirements gathering activity that makes use of facilitated meetings and the development of user scenarios.

Elaboration further expands requirements into an analysis model — a collection of scenario-based, activity-based, class-based, behavioral, and flow-oriented model elements. A variety of modeling notations may be used to create these elements. The model may reference analysis patterns — characteristics of the problem domain that have been seen to reoccur across different applications.

As requirements are identified and the analysis model is created, the software team and other project stakeholders negotiate the priority, availability, and relative cost of each requirement. The intent of this negotiation is to develop a realistic project plan. In addition, each requirement and the analysis model as a whole are validated against customer need to ensure that the right system is to be built.

Dictation

1. engineers
2. follows
3. cases
4. same
5. projects
6. working
7. organizations
8. group
9. originally
10. built
11. overall
12. requirements
13. product
14. department
15. component
16. end-user
17. actually
18. achieve
19. purposes
20. operational

Part 2 Reading and Translating

Section A

Ⅰ. 1. T 2. F 3. T 4. T 5. T

Ⅱ. 1. (B) 2. (D) 3. (C)

Ⅲ.
1. declaration
2. Based
3. expected
4. early
5. relate
6. defines
7. explaining
8. said
9. overcoming
10. faced

Ⅳ. 非功能性需求

非功能性需求有时会根据指标(指能够衡量系统的标准)来定义,以使其更加明确。非功能性需求所描述的也可能不是和系统执行有关的方面,而是与系统随时间产生演变的方面有关(比如可维护性、可扩展性、文档等)。

非功能性需求不是系统的直接需求,而是以某种方式与可用性相关。例如,对于银行应用程序,其主要的非功能性需求是,如果可能的话,应用程序需要满足一周 7 天一天 24 小时都能使用且不中断。

Section B

Ⅰ. 1. F 2. F 3. F 4. T 5. F

Ⅱ. 1. (B) 2. (A) 3. (D)

Ⅲ.
1. capturing
2. environment
3. However
4. through
5. broken
6. delivered
7. stories
8. only
9. As
10. ensures

Ⅳ. UML

UML 是 Unified Modeling Language(统一建模语言)的缩写,是一种由集成图组成的标准化的建模语言,旨在帮助系统和软件开发人员对软件系统的构建、业务建模和其他非软件系统进行详细描述、可视化、构建和记录。UML 在一系列极佳的工程实践中已被证明其能在大型和复杂系统的建模方面取得成功。UML 是开发面向对象软件和软件开发过程的非常重要的工具。UML 主要使用图形符号来表示软件项目的设计。使用 UML 可以帮助项目团队交流,探究潜在的设计方案并验证软件的体系结构设计。

Answers(练习答案)

Unit 3

Part 1 Listening and Speaking

Listening Comprehension

1．（B）　　　2．（D）　　　3．（C）

原文如下：

Software Project Planning

There are many different planning **philosophies**. Some people are "**minimalists**", **arguing** that change often **obviates** the need for a detailed plan. Others are "**traditionalists**", arguing that the plan provides an effective **road map**, and the more detail it has, the less likely the team will become lost. Still others are "**agilists**", arguing that a quick "planning game" may be necessary, but the road map will emerge as "real work" on the software begins.

What to do？On many projects, over-planning is time consuming and **fruitless**（too many things change）, but under-planning is a **recipe** for **chaos**. Like most things in life, planning should be conducted **in moderation**, enough to provide useful guidance for the team — no more, no less.

In an excellent **paper** on software process and projects, Barry Boehm **states**: "You need an organizing principle that **scales down** to provide simple plans for simple projects." Boehm suggests an approach that addresses project objectives, **milestones** and schedules, responsibilities, management and technical approaches, and required resources. He calls it the **W5HH** principle, after a series of questions that lead to a definition of key project characteristic and the **resultant** project plan: Why is the system being developed？What will be done？When will it be accomplished？Those above are the most indispensable. In addition, who is responsible for a function？Where are their organizations located？How will the job be done technically and **managerially**？And how much of each resource is needed？The answers to Boehm's W5HH questions are important regardless of the size or complexity of a software project.

Dictation

1．activities　　2．pre-defined　　3．having　　4．process　　5．schedule
6．classic　　7．control　　8．interacting　　9．over-budget　　10．quality
11．within　　12．lifecycle　　13．goes　　14．baby　　15．visible
16．choose　　17．implied　　18．inputs　　19．pick　　20．meeting

Part 2 Reading and Translating

Section A

Ⅰ．1．T　　2．F　　3．F　　4．T　　5．T

Ⅱ．1．（B）　　2．（C）　　3．（D）

Ⅲ．
1．assembled　　2．scheduled　　3．take　　4．pick　　5．Integrating
6．no　　7．help　　8．chart　　9．through　　10．ahead

Ⅳ．工作量估算的不准确性

顺理成章地，项目策划，尤其是其中工作量评估的重要性和难度会因团队、公司和项目而异。在敏捷策划中，团队通常只策划下一个简短的冲刺（sprint）（甚至可以利用先前的冲刺或已发行版本的反馈），而与其相比，策划某个完整的瀑布型的软件开发生命周期的过程会是一项艰巨的任务。

来自依靠主观判断（无论他们是否是专家）的价格敏感客户的偏见，或者借由单一信息来源而不是多方信息来源来进行估算，以及在多数情况下缺乏历史数据都会对估算准确性带来负面影响。尽管上述风险有一部分可以通过组织性的改变以及修正管理方法来减轻，但修正管理方法还需要使用合适的工具。

只有使用支持这些过程的软件工具，才能有效地收集和组织历史数据，并对其进行分析以提高对未来项目估算的准确性。

Section B

Ⅰ. 1. F   2. T   3. F   4. T   5. F

Ⅱ. 1.（C）   2.（B）   3.（C）

Ⅲ.

1. However   2. equipped   3. Having   4. implementing   5. addition

6. take   7. carefully   8. opportunities   9. facilitating   10. return

Ⅳ. 项目经理和 AI

人工智能将会协助而不是取代项目经理。与所有技术一样，仅凭 AI 不能保证成功。但是通过有目的的部署，AI 可以成为项目经理的"独特推进器"，以及游戏规则的颠覆者，从而有助于提高项目的成功率。成功的项目经理可能会是那些设法超越人类想象力界限的人，他们能够回答 AI 如何增加实际存在的价值这样的问题，以及回答 AI 如何推动项目管理和业务转型以产生积极变化的问题。这将确保项目管理的战略价值。

Unit 4

Part 1　Listening and Speaking

Listening Comprehension

1.（C）   2.（B）   3.（D）

原文如下：

Project Team

    There are many reasons that software projects get into trouble. Many projects in which the scale of efforts is large need to be developed by project teams ranging in size from two to several hundred people. That is leading to complexity, confusion, and significant difficulties in coordinating team members. Putting together a group that works effectively is therefore a critical management task. A good group has a team spirit so that the people involved are **motivated** by the success of the group as well as by their own personal goals.

    Ian Sommerville describes this spirit as "**cohesiveness**" in his book Software Engineering. In a cohesive group, members think of the group as more important than the individual in it. Members of a **well-led**, cohesive group are loyal to the group. They **identify with** group goals and with other group members. They attempt to protect the group, as an entity, from outside interference. This makes the group robust and able to **cope with** problems and unexpected situations. The group can cope with change by providing mutual support and help. The advantages of a cohesive group include a group quality standard that is established by **consensus** and observed by self, group members working closely together, learning from each other, getting to know each other's work, practicing **egoless** programming, and so on.

    In another book, Peopleware, DeMarco and Lister call that spirit as "**jell**". They discuss this issue as follows. A **jelled** team is a group of people so strongly **knit** that the whole is greater than the sum of the parts. Once a team begins to jell, the probability of success **goes way up**. The team can become unstoppable, a **juggernaut** or success… they do not need to be managed in the traditional way, and they certainly do not need to be motivated. They have got **momentum**.

    To strengthen the cohesiveness and jell of a group, some effective methods must be established for coordinating the members to come to understand the motivations, strengths and weaknesses of other people in the group. Good communication between members of a software development group is one of the most essential ways. The group members must exchange information on the status of their work, the design decisions that have been made and changes to previous decisions that are necessary. To accomplish this, mechanisms for formal and informal communication among team members must be established. Formal communication is accomplished through "writing, structured meetings, and other relatively non-interactive and impersonal communication channels". Informal communication is more personal. Members of a software team share ideas on an ad hoc basis, ask for help as problems arise, and interact with one another on a daily basis.

Dictation

1. fast-paced
2. Market
3. evolve
4. emerge
5. enough
6. business
7. Initially
8. lightweight
9. people-centric
10. Alliance
11. statement
12. interactions
13. documentation
14. customer
15. responding
16. prior
17. Extreme
18. Adaptive
19. Feature
20. Dynamic

Part 2 Reading and Translating

Section A

Ⅰ. 1. T   2. T   3. F   4. F   5. T
Ⅱ. 1.（C）   2.（A）   3.（C）
Ⅲ.
1. effective   2. defined   3. hand   4. resolve   5. works
6. critical   7. like   8. enough   9. reaching   10. created

Ⅳ. 如何组建成功的开发团队（1）

优秀的团队不是靠自己形成的,但是为什么我们最终还是需要优秀的团队呢？难道优秀的专业人士不能自发地组建优秀的团队吗？问题在于,仅仅将优秀的专业人士聚集在一起并给他们一个项目的截止日期是不够的。高效的团队不仅和专业性有关,也和团队成员之间的互动方式有关。只有真正高效的团队才能更快地处理工作负荷并更加富有成效。通常而言,和一支井然有序的团队一起工作总会是一件令人愉快的事情。

Section B

Ⅰ. 1. T   2. F   3. T   4. T   5. F
Ⅱ. 1.（D）   2.（C）   3.（D）
Ⅲ.
1. from   2. into   3. on   4. including
5. communication   6. accepted   7. aimed   8. while
9. to   10. preferable

Ⅳ. 如何组建成功的开发团队（2）

统计数据表明,事实上项目失败的主要原因是参与者对项目能否成功信心不足："75%的受访者承认,他们的项目通常从一开始注定就要失败。"但是为什么你认为这样的事情会发生呢？你又该如何改变这样的局面呢？为了正常工作,团队需要了解流程的所有方面,需要了解他们的任务和责任,并相信他们所作的工作——你就是使他们相信这些的人。这就是如何建立一个可以依靠的软件开发团队,一个有信心的高效团队,一个可以看到并说"我们将改变世界"的方法。

Unit 5

Part 1 Listening and Speaking

Listening Comprehension

1.（D）   2.（C）   3.（B）

原文如下：

Software Design

  Design engineering **commences** as the first iteration of requirements engineering **comes to a conclusion**. The intent of software design is to apply a set of principles, concepts, and practices that lead to the development of a high-quality system or product. The goal of design is to create a model of software that will implement all customer requirements correctly and bring delight to those who use it. Design engineers must **sift** through many design alternatives and **converge on** a solution that best suits the needs of project stakeholders.

  The design process moves from a "big picture" view of software to a more narrow view that defines the detail required to implement a system. The process begins by focusing on architecture. Subsystems are defined; communication mechanisms among subsystems are established; components are identified; and a detailed description of each component is developed. In addition, external, internal, and user interfaces are designed.

  Design concepts have evolved over the first half-century of software engineering work. They describe attributes of computer software that should be present regardless of the software engineering process that is chosen, the design methods that are applied, or the programming languages that are used.

  The design model **encompasses** four different elements. As each of these elements is developed, a more complete view of the design evolves. The architectural element uses information derived from the application domain, the analysis model, and available catalogs for patterns and styles to derive a complete structural representation of the software, its subsystems and components. Interface design elements model external and internal interfaces and the user interface. Component-level elements define each of the modules (components) that **populate** the architecture. Finally, deployment-level design elements allocate the architecture, its components, and the interfaces to the physical configuration that will **house** the software.

Dictation

  1. Human-Computer 2. experience 3. rely 4. usable 5. acceptance
  6. key 7. process 8. focus 9. specialists 10. relationship
  11. control 12. blame 13. Contrarily 14. accomplish 15. well-designed
  16. behaves 17. place 18. load 19. consistent 20. golden rules

Part 2 Reading and Translating

Section A

  Ⅰ. 1. T 2. T 3. F 4. T 5. T
  Ⅱ. 1.（D） 2.（D） 3.（C）
  Ⅲ.
  1. Depending 2. look 3. given 4. follow 5. understanding
  6. building 7. writing 8. long 9. compare 10. help
  Ⅳ. 设计复用

  设计复用是通过重用以前开发时的设计来构建新的软件应用程序和工具的过程。可以通过融入较小的变更来添加新的特性和功能。

  设计复用涉及使用所设计的模块（例如，逻辑和数据）来构建改进了的新产品。可重用的组件（包括代码段、结构、计划和报告）可最大程度地减少实施时间，并且成本较低。这样可以通过使用已经研制的技术避免彻底改造现有的软件以创建和测试软件。

Section B

Ⅰ. 1. F　　　2. F　　　3. F　　　4. T　　　5. T
Ⅱ. 1.（D）　　2.（D）　　3.（B）
Ⅲ.
1. serves　　2. using　　3. enables　　4. based　　5. order
6. supported　　7. wrote　　8. meets　　9. short　　10. depicting

Ⅳ. 设计模式

设计模式是软件工程问题的可复验解决方案。与大多数特定于项目的解决方案不同,设计模式在众多项目中都能被使用。设计模式不被视为成品,相反,它们是可以应用于多种情况的模板,并且随着时间的推移可以进行改进,从而使它们成为非常强大的软件工程工具。由于使用成熟的原型可以提高开发速度,因此使用设计模式模板的开发人员可以提高编码效率和最终产品的可读性。

Unit 6

Part 1　Listening and Speaking

Listening Comprehension

1.（A）　　2.（D）　　3.（B）

原文如下：

Writing the Code

So far, the skills we have learned helped us to understand the customers and users problem and to devise a high-level solution for it. Now, we must focus on implementing the solution as software. That is, we must write the programs that implement the design. This task can be **daunting**, for several reasons. First, the designers may not have addressed all of the characteristics of the platform and programming environment; structures and relationships that are easy to describe with charts and tables are not always straightforward to write as code.

Second, we must write our code in a way that is understandable not only to us when we revisit for testing, but also to others as the system evolves over time. In the real world, most software is developed by teams, and a variety of jobs are required to generate a quality product. Even when writing the code itself, many people are usually involved, and a great deal of cooperation and coordination is required. Thus, it is very important for others to be able to understand not only what you have written, but also why you have written it and how it **fits in with** their work. To be a truly excellent programmer, you must know that programming is communicating with other programmers first and communicating with the computer second.

Third, we must **take advantage of** the characteristics of the design's organization, the data's structure, and the programming language's **constructs** while still creating code that is easily reusable. The most critical standard is the need for a direct **correspondence** between the program design components and the program code components. The entire design process is of little value if the design's **modularity** is not carried forward into the code. Design characteristics, such as low **coupling**, high **cohesion**, and well-defined interfaces, should also be program characteristics, so that the **algorithms**, **functions**, interfaces, and data structures can be traced easily from design to code and back again. In case of requirements changing, they are made first to the high-level design and then are traced through lower design levels to the code that must be modified.

Although much of coding is an individual endeavor, all of coding must be done with your team in mind. Your use of information hiding allows you to reveal only the essential information about your components so that your colleagues can invoke them or reuse them. Your use of standards enhances communication among team members. And your use of common design techniques and strategies makes your code easier to test, maintain, and reuse. No matter which programming language to be used, and no matter how the code to be worked out, you should keep some of the software engineering practices above in mind as you write your code.

Dictation

1. effects   2. causes   3. minority   4. greatest   5. quality
6. named   7. economist   8. population   9. Similarly   10. following
11. contribute   12. execution   13. experience   14. program   15. loops
16. doubled   17. hot spots   18. resources   19. percent   20. daily

Part 2 Reading and Translating

Section A

Ⅰ. 1. T   2. F   3. T   4. T   5. F
Ⅱ. 1. (D)   2. (D)   3. (C)
Ⅲ.
1. comes   2. building   3. laid   4. writing   5. understand
6. starts   7. development 8. possibilities 9. converted 10. Based

Ⅳ. 代码审查

代码审查或者同行代码审查是一种有意识、有组织地召集某个程序员的同事来检查彼此的代码是否有错误的行为,并且已被反复证明可以加快和简化软件开发过程,这是其他实践所无法做到的。虽然已经有同行代码审查工具和软件,但是理解概念本身很重要。软件是由人编写的,因此,软件经常充满错误,犯这些错误的当然是人,所以错误显然是与人相关的。然而,不知道为什么软件开发人员经常依靠手动或自动测试来审查他们的代码,而忽略了其他人性的天赋:自己发现并纠正错误的能力。

Section B

Ⅰ. 1. T   2. F   3. F   4. T   5. F
Ⅱ. 1. (D)   2. (D)   3. (C)
Ⅲ.
1. changing   2. covers   3. from   4. throughout   5. without
6. cleaning   7. applied   8. appear   9. maintains   10. plays

Ⅳ. 软件复杂度

Grady Booch(UML 的合著者之一)在他的《面向对象的设计》一书中,在第 1 章中描述了为什么世界原本是复杂的(包括其所有事件及其属性),因此软件必须复杂,因为它代表了现实。但是,Booch 忽略了要说明的一点是,这种复杂性可以很好地被抽象化,从而使最终用户不必处理它。换句话说,Booch 提到的人的抽象能力只是为了将其解释为在面向对象编程范例中建模的大脑过程,但与他第 1 章所述的复杂性没有关系。

简言之,尽管现实原本是复杂的,并且软件是现实的模型,但是可以通过建立可创建新模型的基础框架来简化软件;这一做法可以隐匿大部分的复杂性,并允许较高层的用户仅从字面上关注他们打算解决的问题。

Answers(练习答案)

Unit 7

Part 1 Listening and Speaking

Listening Comprehension

1. (B)   2. (A)   3. (C)

原文如下：

Software Testing

Testing plays a critical role in quality assurance for software. Due to the limitations of the **verification** methods for the previous phases, design and requirement faults also appear in the code. Testing is used to detect these errors, in addition to the errors **introduced** during the coding phase.

Testing is a dynamic method for verification and validation, where the system to be tested is executed and the behavior of the system is observed. Due to this, testing observes the failures of the system, from which the presence of faults can be **deduced**. However, separate activities have to be performed to identify the faults and then remove them.

There are two approaches to testing: black-box and white-box. In black-box testing, the internal logic of the system under testing is not considered and the test cases are decided from the specifications or the requirements. It is often called functional testing. **Equivalence class partitioning, boundary value analysis** and **cause-effect graphing** are examples of methods for selecting test cases for black-box testing. **State-based testing** is another approach in which the system is modeled as a state machine and then this model is used to select test cases using some transition or path based coverage criteria. State-based testing can also be viewed as **grey-box testing** in that it often requires more information than just the requirements.

In **white-box testing**, the test cases are decided entirely on the internal logic of the program or module being tested. The external specifications are not considered. Often a criterion is specified, but the procedure for selecting test cases is left to the tester. **Mutation testing** is an approach for white-box testing that creates **mutants** of the original program by changing the original program. The testing criterion is to kill all the mutants by having the mutant generate a different output from the original program.

As the goal of testing is to detect any errors in the programs, different levels of testing are often used. Unit testing is used to test a module or a small collection of modules and the focus is on detecting coding errors in modules. During integration testing, modules are combined into subsystems, which are then tested. The goal here is to test the system design. In system testing and acceptance testing, the entire system is tested. The goal here is to test the system against the requirements, and to test the requirements themselves. White-box testing can be used for unit testing, while at higher levels mostly black-box testing is used.

Dictation

1. smoke	2. assembly	3. validating	4. checked	5. defects
6. confirm	7. expected	8. says	9. integration	10. build
11. executable	12. exposing	13. benefits	14. critical	15. risk
16. conducting	17. uncovering	18. correction	19. assessment	20. a bit

Part 2 Reading and Translating

Section A

Ⅰ. 1. T   2. F   3. T   4. T   5. T

Ⅱ. 1. (B)   2. (B)   3. (D)

Ⅲ.
1. carried   2. main   3. discovered   4. conducts   5. involves
6. called   7. based   8. met   9. interpreted   10. Commonly

Ⅳ. 常规测试与面向对象测试

常规测试是传统的测试方法，通常在使用瀑布生命周期进行开发时进行测试，而在使用面向对象的分析和设计来开发企业软件时使用面向对象的测试。常规测试更多地关注分解和功能方法，而面向对象的测试则使用组合的方法。常规测试中使用的三个测试级别（系统、集成、单元）对于面向对象的测试来说没有明确的界定。这样做的主要原因是，面向对象的开发使用增量方法，而传统的开发则采用顺序方法。就单元测试来说，与常规测试相比，面向对象的测试所测试的单元看起来更小。

Section B

Ⅰ. 1. F   2. F   3. T   4. T   5. F
Ⅱ. 1.（C）   2.（B）   3.（D）
Ⅲ.
1. which   2. Originally   3. includes   4. considered   5. referred
6. subjected   7. releasing   8. use   9. available   10. making

Ⅳ. 回归测试

每当开发人员更改或修改他们的软件时，即使很小的调整也可能产生意想不到的后果。回归测试是测试现有的软件应用程序，以确保所做的更改或添加未破坏任何现有的功能。其目的是捕获可能意外引入到新的 Build 版或候选发布版中的错误，并确保先前已消除的错误已被消除。通过重新运行最初解决已知问题时编写的测试方案，可以确保对应用程序的任何新的更改都不会导致性能下降或导致以前正常工作的组件失效。此类测试可以在小型项目上手动执行，但是在大多数情况下，每次进行更新时都重复进行一整套测试非常耗时且复杂，因此通常需要使用自动化测试工具。

Unit 8

Part 1 Listening and Speaking

Listening Comprehension

1.（C）   2.（A）   3.（C）

原文如下：

Software Delivery

The delivery of a software increment represents an important milestone for any software project. But before your **cheer** for that, something important must be completed. First of all, you must ensure that your delivery should have been tested and fixed without bugs. Second, you should deliver more than the software itself. That is to say, besides executable software, support data files, support documents, and other relevant information must be assembled together. An appropriate instructional materials must be provided to end-user such as appropriate training aids, troubleshooting guidelines, and a "what's-different-about-this-software-increment" description. Furthermore, a support **regime** must be established before the software is delivered. An end-user expects responsiveness and accurate information when a question or problem arises. If support is ad hoc, or worse, nonexistent, the customer will become dissatisfied immediately. Support should be planned, support material should be prepared, and appropriate record keeping mechanisms should be established so that your software team can conduct a **categorical** assessment of the kinds of support requested.

The delivered software provides benefit for the end-user, but it also provides useful feedback for the software team. As the increment is **put into use**, the end-users should be encouraged to comment on features and functions, ease of use, reliability, and any other characteristics that are appropriate. Feedback should be collected and recorded by the software team and used to make immediate modifications to the delivered increment, define changes to be incorporated into the next planned increment, make necessary design modifications to accommodate changes, and revise the plan for the next increment to reflect the changes.

Dictation

1. popularly   2. observed   3. defects   4. errors   5. entering
6. indication   7. external   8. internal   9. relationship   10. connects
11. locating   12. device   13. commercial   14. application   15. portion
16. fail   17. quit   18. divide   19. trail   20. fresh

Part 2　Reading and Translating

Section A

Ⅰ. 1. T　　2. F　　3. T　　4. T　　5. T
Ⅱ. 1. (C)　　2. (A)　　3. (C)
Ⅲ.
1. Whether   2. involves   3. Besides   4. specific   5. as
6. Unfortunately　7. cases   8. means   9. first   10. on

Ⅳ. 软件维护的类型

软件维护有四种类型，即纠正性、适应性、完善性和预防性维护。纠正性维护与在使用软件时观察到的修复错误有关；适应性维护与为了使软件适应新环境（例如，在新操作系统上运行软件）而发生的软件更改有关；完善性维护与在软件中添加新功能时发生的软件更改有关；预防性维护涉及实施更改以防止错误发生。软件维护的种类按类型和所用时间百分比来确定。

Section B

Ⅰ. 1. T　　2. F　　3. F　　4. F　　5. T
Ⅱ. 1. (D)　　2. (B)　　3. (D)
Ⅲ.
1. depends   2. known   3. seen   4. respect   5. deals
6. include   7. accessed   8. done   9. match   10. terms

Ⅳ. 软件工程质量

从广义上讲，软件质量定义中"针对开发目的的适应性"是指对需求的满足。但是有哪些需求？需求在当今的敏捷术语中也称为用户故事，可以分为功能性和非功能性两种。功能需求是指软件应能够实现的特定功能。例如，在 HP M1136 打印机上进行打印的功能是一项功能需求。但是，仅由于软件具有某种功能或用户可以使用该软件完成某个任务，并不意味着该软件具有良好的质量。可能在很多情况下，你使用过软件并达到了预期的目的，例如，找到航班或预订酒店，但你认为软件质量很差。这是因为实现功能的"方式"。对如何实现功能不满意表示软件未满足非功能性需求。

Unit 9

Part 1　Listening and Speaking

Listening Comprehension

1. (B)   2. (A)   3. (C)

原文如下：

Expansion Company

  As a part of job invitation advertisement, here is an overview on Expansion, a company providing software products and services in IT industry:

  Expansion is one of the top software products, services and solutions providers in China. It was established in 2010, and now is a leading IT company in China, but also has a global presence. It headquarters in the heart of China's Silicon Valley, Zhong Guancun, Beijing. Expansion helps to meet the outsourcing needs of companies worldwide, and its goal is to become a global leading IT outsourcing service company by providing quality services to global customers.

  In the early years of Expansion, testing service was its major business. With the development of the company, there are more and more various products and services to provide today. Among them, Software Development and Application Maintenance Service have become the key business. In addition, BPO, shorted for Business Process Outsourcing, is a kind of rising service which Expansion begins trying.

  In the past decade, Expansion has a brilliant track record as an IT partner for Fortune 500 companies, including Microsoft, Google, IBM, and so on. Japan is the main customer among its overseas market at present. Meanwhile, Europe and U.S. will become another two huge markets shortly by establishing its office there.

  Expansion has an experienced management team and a rich pool of well educated, well trained and highly experienced engineers and project managers. Further, its staffs are expected to grow over 900 members in 2025. Through the endeavor from all of them, Expansion has been keeping margin increase at the rate of over 10% every year since foundation.

  In the future, Expansion will attract new clients continually to develop global presence, advance strong management team and human resources by investing in employees' trainings to develop their experience, and become more profitable by high efficiency and top quality in work. By all of those efforts, Expansion will be very hopeful to realize its strategic goal.

Dictation

  1. change   2. effectively   3. control   4. likely   5. establishing
  6. versions   7. reporting   8. eventually   9. detect   10. integration
  11. requirements   12. documentation   13. idea   14. kept   15. repository
  16. copies   17. back   18. track   19. features   20. versioning

Part 2   Reading and Translating

Section A

  Ⅰ. 1. T   2. F   3. F   4. F   5. T
  Ⅱ. 1. (B)   2. (D)   3. (A)
  Ⅲ.
  1. intended   2. releases   3. combines   4. knowledge   5. begins
  6. request   7. designed   8. which   9. accepted   10. deployed
  Ⅳ. 软件架构

  软件架构是一种结构性的框架，用于使软件元素、关系和属性概念化。软件架构还引用了软件体系结构文档，该文档有助于与利益相关者进行交流，同时记录有关不同项目的设计和组件以及模式重用的早期的和高层的决策。可通过对重要事情的抽象和分离来降低软件架构过程的复杂性。

Section B

  Ⅰ. 1. T   2. T   3. F   4. F   5. T

Ⅱ. 1.（D）   2.（D）   3.（A）

Ⅲ.

1. used	2. collected	3. helps	4. In	5. namely
6. like	7. reported	8. many	9. as	10. to

Ⅳ. 软件质量保证（SQA）

SQA 能够帮助确保开发高质量的软件。不论使用哪种底层的软件开发模型，大多数类型的软件开发都采用 SQA 实践。从广义上讲，SQA 包含并采用软件测试方法来测试软件。SQA 不会在开发完成后检查质量，而是在开发的每个阶段都对质量进行测试，直到软件完成为止。使用 SQA，仅当当前/上一个阶段符合所需的质量标准时，软件开发过程才能进入下一阶段。

SQA 通常根据一种或多种行业标准开展工作，这些标准有助于制定软件质量准则和执行战略。这些标准包括 ISO 9000 和能力成熟度模型集成（CMMI）。

Unit 10

Part 1 Listening and Speaking

Listening Comprehension

1.（A）   2.（B）   3.（A）

原文如下：

The Organizational Structure of a Company

It is important that every new member gets to know the organizational structure of his company and realizes the relationships of membership as soon as possible. The brief description below about the company organization will facilitate the understanding of the company that a new employee works for.

In a company, the managers and **subordinates** make up an organization. There are some main **chiefs taking the lead**. The first one is CEO, which is shortened form of Chief Executive Officer. He is the executive who is responsible for a company's overall **operations**, usually is the President or/and the Chairman of the **Board** of a company. In addition, Chief Financial Officer, he is senior manager who is responsible for **overseeing** the financial activities of an entire company, including financial planning and cash flow, **CFO is for short**. Another senior manager is CTO, whose full name is Chief Technology Officer. He is in charge of technology related departments in the company such as department of IT Support and department of Information System. Besides, **COO** is the **acronym** of Chief Operating Officer, who is senior manager whose responsibility is to manage the day-to-day operations of the company and then report this to the CEO. They form the senior leadership of the company.

The organization mentioned above is **classed** in terms of **rank**. So there is a hierarchical structure in a company: the CEO is the head of the company; the CTO, CFO and COO report directly to the CEO; the Director of Marketing and the Manager of Customer Services are **under the supervision** of the COO. As this company grows, its organization may become much larger and the real structure is likely to be much more complex. That is to say, every position in this structure can be expandable, maybe the CTO and the CFO may each have teams working under them. In some companies, there are **CBDMO**, the Chief Business Development and Marketing Officer, and **CSO**, the Chief Strategy Officer **besides**.

Dictation

1. bundled	2. source	3. business	4. secret	5. distributed
6. harnesses	7. reliability	8. acccss	9. Characterized	10. birth
11. held	12. announcement	13. release	14. innovative	15. operating systems
16. generated	17. challenge	18. exception	19. procedures	20. mechanisms

Part 2　Reading and Translating

Section A

Ⅰ. 1. T　　2. F　　3. T　　4. F　　5. T

Ⅱ. 1.（A）　2.（C）　3.（D）

Ⅲ.

1. discussing　2. concerned　3. undergoes　4. Given　5. updating
6. affecting　7. done　8. required　9. far　10. existing

Ⅳ. 良好沟通的重要性

　　有意思的是，在软件开发中拥有良好的交流方法和技能确实至关重要。对于开发人员、产品和业务而言，传递有关规范、需求、代码中的错误等信息至关重要。提出需求者与实际工作者之间必须要有强烈的共识，否则可能会对手头的问题得到不好的甚至完全错误的解决方案。

　　无论使用哪种沟通方法，很多时候信息都会丢失。为确保人们了解所有必要的信息，你应该向他们提出问题或以其他方式对他们进行测试。这可能很困难，尤其是在受众很多的情况下。

Section B

Ⅰ. 1. F　　2. F　　3. T　　4. T　　5. F

Ⅱ. 1.（D）　2.（B）　3.（D）

Ⅲ.

1. talk　2. short　3. process　4. speed　5. only
6. discovered　7. because　8. through　9. enables　10. focus

Ⅳ. CMMI

　　能力成熟度模型集成（CMMI）是一种过程和行为模型，可帮助组织机构精简过程改进并鼓励富有成效和高效的行为，从而降低软件、产品和服务开发的风险。

　　CMMI 是连续的、有组织的模型。因此，软件组织机构应选择任意一种模型，以及选择一系列行为准则作为模型的一部分，以评估和改进软件过程。

图书资源支持

感谢您一直以来对清华版图书的支持和爱护。为了配合本书的使用,本书提供配套的资源,有需求的读者请扫描下方的"书圈"微信公众号二维码,在图书专区下载,也可以拨打电话或发送电子邮件咨询。

如果您在使用本书的过程中遇到了什么问题,或者有相关图书出版计划,也请您发邮件告诉我们,以便我们更好地为您服务。

我们的联系方式:

地　　址:北京市海淀区双清路学研大厦A座714

邮　　编:100084

电　　话:010-83470236　010-83470237

客服邮箱:2301891038@qq.com

QQ:2301891038(请写明您的单位和姓名)

资源下载: 关注公众号"书圈"下载配套资源。

书圈

获取最新书目

观看课程直播